The Mandela Effect
and its Society

Also by Cynthia Sue Larson

Aura Advantage
Karen Kimball & the Dream Weaver's Web
Quantum Jumps
RealityShifters Guide to High Energy Money
Reality Shifts
Shine with the Aura of Success

The Mandela Effect
and its Society

Awakening from ME to WE

CYNTHIA SUE LARSON

THE MANDELA EFFECT AND ITS SOCIETY
Awakening from ME to WE

It is easy, and perhaps even lazy, to assume that reports of the supposed Mandela Effect can all be attributed to faulty memory. Cynthia Sue Larson makes a convincing case that something deeper is at work. Is it a question of the many worlds interpretation of quantum theory? Is it a matter of multiple timelines intersecting in strange ways? We certainly have more questions than answers. However, one thing is clear to me. We make a serious cognitive error when we refuse to even consider a topic because it seems too bizarre for comprehension. Such anomalies are, in my view, the most important of all.
— Jeffrey Mishlove, PhD Host, *New Thinking Allowed* YouTube channel

The Mandela Effect and Its Society takes readers on a captivating journey into the peculiar phenomenon of clear memories of things that never were. While easy to dismiss these mismatched memories as tricks of the mind, Cynthia Sue Larson's analysis of such episodes in her own life, and her speculations about them, raise intriguing questions about our understanding of reality. Prepare to be captivated as you explore the intriguing web of high strangeness known as the "Mandela Effect."
— Dean Radin PhD, Chief Scientist, Institute of Noetic Sciences, and author, *Real Magic,* and other books

THE MANDELA EFFECT AND ITS SOCIETY
Awakening from ME to WE

The Mandela Effect and Its Society is an extraordinary and illuminating book. Its powerful, profound and alchemical quest to the heart and soul of the human psyche offers a revelatory perspective of the potential for our conscious evolution.
— Dr. Jude Currivan, cosmologist, planetary healer, futurist, author and co-founder of WholeWorld-View

One of the leading researchers of reality shifts and the phenomenon of collective alternate memories known as the Mandela Effect, Cynthia Sue Larson, has given birth to a new and important book. This is a one of a kind brilliant and comprehensive book on the Mandela Effect that only Cynthia, a prolific "in-between worlds traveler," could have authored. There are as many ways to look at time and reality as there are beings on the planet—but Larson has devoted more time and more creative brain power to the subject than just about anyone I know. If you don't already know about the Mandela Effect, now you'll know. Or perhaps you will find out when this book, destined to be a groundbreaking motion picture, comes out. To understand the genius of this book, you must accept the premise that rational thought, once considered the gold standard of thinking, is not what it used to be. Our rational mind is melting down before our eyes so that a new, time-free, multi-dimensional way of being in the world can be born. For some, like Larson, it is already here. Read this book with an open mind, and be prepared to have your reality shattered, only in a fun, playful way with Cynthia as your guide.
— Glenn Aparicio Parry, two-time Nautilus award-winning author of *Original Thinking: A Radical ReVisioning of Time, Humanity and Nature* and *Original Politics: Making America Sacred Again.*

THE MANDELA EFFECT AND ITS SOCIETY
Awakening from ME to WE

The fact of the Mandela Effect is overwhelming when we consider yet the simplest viewing of an event by several people. Even the time of the event or often the location of it is often recalled differently by different observers. From a quantum physical point of view this may not be as surprising as it may seem to non-scientists. Here, Ms. Larson has provided us with a well-researched and engaging book describing this often baffling manifestation. If you have ever experienced the effect you will find this very worth your reading.
— Fred Alan Wolf, author-physicist of many books including *Time-loops and Space-twists: How God Created the Universe.*

Cynthia's new book, *The Mandela Effect and Its Society,* is an extraordinary contribution bringing much needed new light to a controversial and often misunderstood phenomenon. Moving way beyond theories of misremembering and confabulation, Cynthia presents solid science and grounded theories illuminating the profound significance and true meaning of The Mandela Effect. Ultimately, like The Mandela Effect itself, she invites the reader into an evolution of consciousness, at both individual and societal levels, further contributing to Humanity's Great Awakening.
— Roger Kenneth Marsh, author *NexGen Human* and *TruthBubble*

THE MANDELA EFFECT AND ITS SOCIETY
Awakening from ME to WE

Research into consciousness takes another step forward in Cynthia Larson's book, *The Mandela Effect and Its Society*. From Haas avocados and widows who eat fish, to touching on topics like nonlinear perception, double memories, timeline collapse, and bilocation, Larson makes clear that consciousness is primary and fundamental, and that reality shifts dynamically and immediately as consciousness changes.
— Penny Kelly, N.D., author of many books including *Getting Well Naturally, The Evolving Human,* and *Consciousness and Energy.*

The highly gifted Cynthia Sue Larson has given us yet another jewel, her new book, *The Mandela Effect and its Society from ME to WE*. This book is a comprehensive look at and analysis of the Mandela Effect, a fairly recently recognized yet relatively unknown phenomenon. Ms. Larson allows the reader to understand what the Mandela Effect is, its history, examples of its occurrences, some of her own experiences with it, the science behind and explaining it, its implications for society as a whole, how it may represent a trigger for shifts in consciousness and the paradigm shift (or entering the Quantum Age), and a vision for where we may be headed as a species due to its existence. She further takes the experience of reading her book from the theoretical and anecdotal to a participatory one, by including exercises and recommendations for the reader. Ms. Larson's beautifully developed left-brain analytical abilities and extensive knowledge of quantum physics, coupled with her innate spiritual awareness and higher level perception and perspective, give this work a lovely holistic rendering, with the potential of triggering and facilitating readers on more than one level. I feel that *The Mandela Effect and its Society* is destined to become the go-to book on this phenomenon.
— Diane Brandon, author of *Born Aware, Intuition for Beginners, Dream Interpretation for Beginners,* and *Invisible Blueprints*

THE MANDELA EFFECT AND ITS SOCIETY
Awakening from ME to WE

Printing History: First trade paper edition 2024

RealityShifters®
P.O. Box 7393
Berkeley, CA 94707-7393
www.realityshifters.com

ISBN-13: 9798853380028

CONTENTS

DEDICATION

Dedicated to each and every one of us,
all our relations,
and Divine Source.

ACKNOWLEDGEMENTS

My thanks to Mandela Effect experiencers everywhere, and to South African President Nelson Rolihlahla Mandela for his role in bringing us together. Thanks to those who have inspired me to pursue research into consciousness and mind-matter interaction since the beginning of my journey, including: Dan Moonhawk Alford, PMH Atwater, Cleve Backster, Zorigtbaatar Banzar, Deborah Beauvais, Thomas Brewer, Matthew Bronson, Jerome Busemeyer, John Cramer, Larry Dossey, Walter Jackson Freeman III, Barry Harmon, Eva Herr, Stanley Krippner, Bruce Lipton, Leroy Little Bear, Bernard Beitman, JohnJoe McFadden, W. Robynne McWayne, Hamish Miller, Jeffrey Mishlove, Edgar Mitchell, Julia Mossbridge, Jacob Needleman, Sean O'Nuallain, Glenn Aparicio Parry, Anthony Peake, Dean Radin, Ripon Rabbit, Steve Rother, Leon Secatero, Shawn Secatero, Rupert Sheldrake, Henry Stapp, In Hyuk Suh, Russell Targ, Giuseppe Vitiello, George Weissmann, and Fred Alan Wolf. Thanks also to authors and speakers furthering discussion about the Mandela Effect, including: Christopher Anatra, Mary Rose Barrington, Sean Bond, Alexis Brooks, Fiona Broome, Tray Caladan, Eileen Colts, Candace Craw-Goldman, Nicole DeMario, Dale DuFay, Anthony J. Duran, Jan Engels-Smith, Stasha Eriksen, Tony Jinks, Jerry Hicks, Melinda Iverson Inn, Brian MacFarlane, Akronos Mago, Evan Matraia, Regina Meredith, Eva Nie, Paulo M. Pinto, Kimberly-Lynn Hanson, Shane C. Robinson, Anthony Santosusso, Rob Shelsky, Carter Tweed, Vannessa VA, and Rizwan Virk. Very special thanks to all my friends and family.

"There is no passion to be found playing small
in settling for a life that is less than
the one you are capable of living."

—Nelson Mandela

Chapter 1

INTRODUCTION TO THE MANDELA EFFECT

Could immediate evolutionary improvements, instantaneous geographic and anatomical changes, and miraculous spontaneous healing—including recovery from death—be providing us with clues to essential technology, just when we need it most? At this time when a phenomenon of collective alternate memories known as the *Mandela Effect* is gaining popular interest, people are noticing changes to: movies, books, music, food, clothing, animals, anatomy, geography, astronomy, history, and every aspect of life. With people sharing specifically different memories than official history describes, many questions now arise. Why is the Mandela Effect happening? What do experiencers of the Mandela Effect have in common? Do versions of us exist in parallel universes? How can we know if we are experiencing different timelines? Where is the Mandela Effect taking us?

The term "Mandela Effect" became a hot conversational topic when many people thought Nelson Mandela had died in the 1980s, and were surprised to discover that he was still alive. If this was the only instance of the "Mandela Effect," there would be little further interest—but the term stuck, because there have been many more. Some of the best-known Mandela Effects include different memories for such things as: The Berenst_in Bears; the witch in the Disney animated movie, *Snow White*, saying, "_____ *mirror on the wall*"; the physical location of the kidneys and heart in our bodies; Pokémon character Pikachu's tail; and passages in the Bible regarding what animal lies down with the lamb.

Despite the common assumption that Mandela Effect experiencers suffer from "false memories" "collective misremembering," or "confabulation," it's possible something else is going on. Many Mandela Effect changes are consistently observed by large groups of people sharing the same memories that is noticeably different than what has supposedly "always been true."

Like falling in love or feeling intense grief, the Mandela Effect is something one has to experience—and even then, it can take more than one or two experiences before we start to question our foundational assumption that facts and historical events don't change.

Over the past twenty years, interest in the Mandela Effect topic has steadily grown, to the point it is starting to go mainstream. The Mandela Effect has been featured in numerous books, magazine and newspaper articles, TV shows, movies, and conferences. Dozens of internet groups exist on social media where people share thoughts, feelings, experiences, and ideas about this phenomenon.

The Mandela Effect is known by a variety of different names, including: the Mandela Effect, Alternate Histories, Alternate Memories, Glitch in the Matrix, the M Effect, the Quantum Effect, Mass Memory Discrepancy Effect, Real-Life Retroactive Continuity, Retconned, or just the Effect. A few other names for this phenomenon include: Retcon Effect (for Real-Life Retroactive Continuity), reality shifts, and "JOTT" for "just one of those things." We might also call it "alternative recollections," as a descriptive, unbiased term.

We expect to encounter some uncertainty in our memories, but the Mandela Effect experience feels different from typical momentary confusion or doubt. Mandela Effects are observed when at some point, a particular memory is noticed as being so different from what is currently true that people feel a sense of disorientation and surprise. Not only do we find our memories are incorrect—we also may discover the appearance of a completely consistent back-story associated with new-to-us "historical facts."

Thanks to the internet and technological advances, humanity has reached a point where people noticing differences between their memories and historical facts can learn they are not alone, discuss what they observe, and ask questions about what may be going on.

THE MANDELA EFFECT IS BORN

> *"It is in your hands to make a better world for all who live in it."*
>
> —*Nelson Mandela*

The Mandela Effect is named after Nelson Mandela, who many people remember having died while incarcerated on Robbins Island. The first recorded instance of this collective alternate memory happened in 2001 on Art Bell's "Coast-to-Coast AM" late-night radio program [1-1], described as a show that *"attracted an audience of millions of loyal insomniacs."* [1-2] In October 2015, Art Bell talked with me on his *Midnight in the Desert* radio show about reality shifts and what Bell called "the Mandela Effect," and how school textbooks had mentioned Nelson Mandela having died in prison many years earlier. [1-3] Art Bell asked me whether I would call Nelson Mandela being remembered as having died as the Mandela Effect, and I replied that I'd call that particular type of reality shift "Alive Again." Art discussed with a caller about how she also remembered Nelson Mandela having died. Art Bell replied, *"Yeah, I know. I'm getting so many emails and faxes."*

Eight years after Art Bell first covered the topic on late-night radio, author Fiona Broome popularized the term, Mandela Effect, to describe the phenomenon in which groups of people shared a different memory of specific historical events. Broome points out that the term, Mandela Effect, is not a theory, nor does it imply that what is being observed are "false memories," media mistakes, or simple confusions. Broome initially welcomed people to post their Mandela Effect experiences directly on her website, enjoying those early days, *"when our related conversations were fun. It was speculative. Very sci-fi. Thoroughly geeky, and often hilarious. We talked about quantum theories, and referenced holodecks, Star Trek episodes, Sliders (TV series), and so on. Then people discovered that their memories of the Berenstein Bears books weren't quite correct; the books were about the Berenstain Bears. And then, the Mandela Effect topic exploded."* [1-4] Once international news coverage made the Mandela Effect a media darling, Broome stepped back from the din when, in 2022, the Mandela Effect was conflated with being some kind of "conspiracy theory," rather than a fun exploration of the nature of reality.

Some noteworthy evidence of Nelson Mandela's death in the 1980s includes recollection by international broadcasting journalist, Eileen Colts, of having interviewed Nelson Mandela in the mid 1980s, and then having covered his death in 1987. Colts said she was stunned in 1995 to see Nelson Mandela appearing at a large public event in London, England. [1-5] [1-6] Additional residual evidence of Nelson Mandela's passing in the 1980s appears in an episode of the TV show, *Head of the Class,* which mentions a Mandela memorial stadium in, "The 21st Century News" episode that first aired on May 4, 1988. The character, Alan, says, *"Athletes from all nations began arriving in South Africa today, in preparation for the opening ceremonies of the summer Olympics at Nelson Mandela Memorial Stadium."* It's truly stunning to me that this TV show episode provides a kind of time marker by which we get confirmation of the 1987 date provided by Eileen Colts. [1-7]

The Mandela Effect phenomenon invites us to pay closer attention to what is happening around us, and what has changed. Like a real world game of "spot the difference," we can review how we remember songs, books, movies, celebrities, TV shows, and news events—and compare our memories to historical facts.

The topic of Mandela Effects and reality shifts is deeply meaningful to me, and I have experienced it throughout my life, which might be related to my passion for exploring the nature of reality. I first saw the topic of reality shifts described in print in PMH Atwater's 1999 book, *Future Memory,* describing unusual events that near-death experiencers commonly observe. [1-8] I have been reporting on individual recollections of alternate histories since 1999, in the earliest edition of my first book on this subject, *Reality Shifts,* and in my free monthly *RealityShifters* ezine. I explore this topic in interviews with experts in the fields of quantum science such as JohnJoe McFadden, Jerome Busemeyer, and Yasunori Nomura on my blog and *Living the Quantum Dream* radio show. The Mandela Effect phenomenon appears to be intrinsic to the nature of reality, yet so subtle that until quantum science arrived, it was invisible to us.

My primary objective when I started the *RealityShifters* website and ezine in 1999 was to research and share information about alternate histories and reality shifts reported by people around the world. *RealityShifters* provided a safe haven where people's first-hand accounts were respected and documented, with reports from experiencers of sudden, inexplicable changes to people, places, and things. American physicist Dr. Fred Alan Wolf, conveyed this sense of gratitude for finding an explanation for an increasingly common experience as part of his endorsement for my book, *Reality Shifts:*

> *"No, Martha, you are not going crazy,*
> *just witnessing the reality shift around you."* [1-9]

In October 1999, I began documenting first-hand reports of reality shifts on the www.RealityShifters.com website. This documentation includes what kind of shift occurred, who experienced it, when it occurred, and where it was reported. I often converse with people who do not speak English as their first language, and who sometimes struggle to convey the magnitude of their experience, so each first-hand report is edited and clarified as I inquire what each experiencer was thinking, feeling, and doing before, during, and after each event.

Prior to creating the realityshifters website in 1999, I could not find a single website mentioning the term, *reality shifts,* nor sharing experiences of witnessing radical changes to physical reality. These reports included instances of: things appearing, disappearing, transporting, transforming, and changes in the experience of time. Most reports involve apparent alteration of recorded history, such that the only "proof" or evidence that shifts have occurred are based on what was remembered. All documented reality shift experiences are viewable on the original *RealityShifters* website in the "News" section, and in over 200 consecutive pages in "Your RealityShifter Stories." [1-10]

RISE OF THE MANDELA EFFECT

The term Mandela Effect has trended upward since July 2015, as many reporters covering this phenomena are experiencing it, too. *Lewiston Sun Journal's* Mark LaFlamme noticed a change on which side of his father he's standing by in an old family photo [1-11]. *New Zealand Herald's* Karl Puschmann felt certain that the popular children's books ought to be spelled "Berenstein" and not "Berenstain" [1-12]. *San Diego City Beat's* Tom Siebert noticed a change to a memorable scene in a James Bond movie, "Moonraker," in which the girl with the braces who fell in love with another character with shining teeth no longer has braces. [1-13] I watched The James Bond film *Moonraker* several times when I was in college, and I clearly remember the scene where a blonde girl named "Dolly" wearing glasses and braids shared a cheesy moment with Richard Kiels' character, "Jaws," where they looked at one another, noticed their shiny, metallic teeth matched, and fell in love. Now, all versions of this movie show that Dolly doesn't have any braces on her teeth, and supposedly never did.

If we considered each of these cases separately, a single person feeling unsettled by noticing such changes could be chalked up to their having made some kind of cognitive error. When so many people share highly specific alternate memories which differ from historical facts, we can begin to question what's going on.

Thanks to the phenomenal growth of the internet, it is now possible for people around the world to share their experiences of noticing alternate histories, and see if others have seen something similar, too. When I noticed things changing throughout my life, such as when I heard a song being played on the radio in the 1970s "for the first time" that I had already heard so many times, I could not run a quick check on the internet to see if anyone else noticed the same thing, because the internet did not yet exist.

CEO ACKNOWLEDGES MANDELA EFFECT

In 2019, the president of food distribution software company NECS, Christopher Anatra, announced that he and some of his employees and customers had been noticing Mandela Effects for a couple of years. Anatra's programming team received numerous customer reports that their data about food products seemed to be full of errors. Clients noticed that a certain customer of theirs has a standing order of "X" amount of one kind of food product each month, but now their NECS data shows that they've "always" purchased "Y" amount. Additionally, some food items appeared different in the computer system than what stores remembered they had always ordered in the past, such as "Hass" avocados or "Kraft Stove Top Stuffing," instead of "Haas" avocados and "Stouffer's Stove Top Stuffing" that they remembered purchasing.

These types of errors are not the typical type of computer errors programmers are accustomed to seeing, where computer bugs can be hunted down and corrected in the code. When Anatra's team diligently investigated these glitches, the team was surprised to see that not only did the inventory numbers add up and total properly, but the brand names appeared to have been retroactively changed as well. Chris, like his customers, also remembered "Haas" avocados and Stouffer's Stovetop Stuffing, but computer records showed something different. [1-14]

I find it fascinating that Christopher Anatra suddenly noticed a large yellow sundial located prominently on the side of a house he drove past on his daily drive for more than 20 years. Further investigation about the timing of the sundial's sudden appearance led to the discovery that it had "always" been there. [1-15] There is beautiful synchronicity here, since I witnessed a sundial suddenly appearing one day in the 1990s when I was on a walk with two friends. That sundial was new to all three of us, and showed up in front of us when I raised the topic of reality shifts! [1-9] I love the symbolism of sundials suddenly appearing, since they represent our attempt to track time as we move through space, and our witnessing the arrival of new-to-us sundials seems to be a beautiful invitation to relate to time and space from a whole new perspective and mental framework.

Anatra is unconcerned about backlash from clients or bad publicity, since the quality of his company's software and expertise speaks for itself. Anatra felt motivated to publicly acknowledge the reality of the Mandela Effect, as Matthew Harris explains:

> *"as someone in his position who has seen what he has, he was obligated to share his experiences so that others may stop thinking that they simply remembered something wrong."* [1-15]

IDENTIFYING GENUINE MANDELA EFFECTS

While the Mandela Effect demonstrates that facts and histories can change, not everyone remembers things the same way. I've received emails from people writing, "I think I've experienced a Mandela Effect, but I'm not sure," indicating they would love to have more confidence in recognizing genuine Mandela Effects. There is understandably some confusion on this matter, likely driven by thinking along the lines popularized by French mathematician Pierre-Simon Laplace that,

"The weight of evidence for an extraordinary claim must be proportioned to its strangeness."

One reason that controversy exists regarding the Mandela Effect is that its very nature flies in the face of our assumption of objective reality. If all Mandela Effect experiencers were in complete agreement about which changes to official facts and histories were authentic, proving the reality of this highly subjective personal phenomenon would be easier.

Yet, experiencers are most often convinced by personal memories of inexplicably impossible events that they know happened in their own lives. These might include such things as seeing an old family photograph looking completely different, or items disappearing and reappearing. Such personal proof can often be credible to experiencers, but may not be quite so persuasive to others.

HOW CAN WE KNOW THE DIFFERENCE?

How can we discern whether we've witnessed an actual Mandela Effect, or are simply forgetful or confused? Three simple methods you can employ to discern whether something is a genuine Mandela Effect are to: find anchor memories, look for flip-flops, and confirm with others.

(1) FIND ANCHOR "PROOF" IN MEMORIES

Like wilderness travelers advised to *"take only memories, leave only footprints,"* explorers of parallel worlds appear bound by a similar arrangement. While we'd love to be able to present indisputable evidence capable of convincing anyone that we truly remember something completely different than "how it's always been," in most cases, our memories are our main evidence of Mandela Effects. Anchor memories are specific aspects of a particular memory that help us confirm that we are correctly remembering—rather than misremembering—something, such as remembering having laughed at the corny scene in the James Bond movie, *Moonraker,* where Dolly and

Jaws smile metal-toothed grins at one another, because we remember that Dolly wore braces.

(2) LOOK FOR FLIP-FLOPS

While it is sometimes possible to physically witness more than one reality, such observations typically don't happen at the same time, and evidence for each separate reality seldom appears concurrently. That being said, genuine Mandela Effects can sometimes be experienced as "flip-flopping" back and forth between different possible realities.

I noticed a Mandela Effect flip-flop involving the Costco store name changing from Costco to Cosco for a few weeks, and then back to Costco. I've also witnessed some personal Mandela Effect flip-flops including: business hours changing back and forth between "always being open until 10 pm" or "always being open until midnight." I've also seen the rain gutters on my neighbor's house going from having had leaf guards, or never having had leaf guards. And I witnessed our elderly dog switching back and forth between developing visible cataracts in both eyes, and not having cataracts.

Flip-flops are noteworthy precisely because we typically start to pay a great deal more attention to what is switching back and forth, rather than less attention, and we can see reality shifts even with such focused, intense observation.

(3) CONFIRM WITH OTHERS

Whenever possible, confirm what you remember with others. You can obtain optimal validation for shared experiences of Mandela Effects by asking open-ended questions that don't "lead" toward any particular details. By asking someone what they recall, with minimal information included in the question, it is possible to hear unbiased answers, rather than the answer we're hoping to hear. An interesting aspect of the Mandela Effect is that when our attention is riveted on a possible case of Mandela Effect, we might not come across physical proof or evidence showing other realities. For example, if we recall that the song on the TV show, *Mister Rogers Neighborhood,* includes him singing, "It's a beautiful day in *the* neighborhood," which is different from the way it now seems to be, "It's a beautiful day in *this* neighborhood," we will not hear both versions of the song on the show, but rather just one version. Genuine examples of the Mandela Effect can sometimes be validated by confirming that we cannot turn up side-by-side examples of, for example, a corporate logo in transition—whose existence would tend to indicate we are not viewing a true example of the Mandela Effect, but rather a more mundane and perfectly normal evolving-through-time logo.

QUANTUM PARADIGM

Modern science functioned fairly smoothly for the last few hundred years, until it tripped over the so-called "hard problem of consciousness," [1-16] when scientists realized the existence of a hitherto unseen blind spot within the assumptions of material realism. [1-17] The "hard problem of consciousness" has to do with the way consciousness does not lend itself to being completely understood purely with computational or neural mechanisms. Clearly, there is something elusive yet unmistakably present in what it means to be conscious that is not part of any scientific measurement. With the arrival of a new quantum paradigm, we clearly need to revise both our scientific methodology, and our assumptions about the cosmos and our role in it. While most people remain blissfully unaware of these mind-boggling developments, some people receive such news with blessed relief. Mandela Effect experiencers are aware that long-standing assumptions about reality are untrue, and have more questions about the nature of reality.

Quantum physics shows us that we need to revise both our scientific methodology and our assumptions about reality. Some defunct assumptions include: matter is not all there is, nonlocal influence and effects do occur, objectivity cannot be assumed, and at best, we can only ever know information probabilistically. [1-18]

We are experiencing a kind of "quantum invasion" as every branch of science now includes elements and influence of quantum physics. When we consider the Mandela Effect in the context of this dawning of the new Quantum Age, we see that we might call such things "alternate recollections," with awareness of the idea that each of us exists in a superimposed state from which we are capable of accessing many possible histories, present moments, and futures. The idea that the many worlds of quantum physics might be one and the same as the multiverse has been proposed by esteemed physicists Dr. Yasunori Nomura and Dr. Raphael Bousso of UC Berkeley, while increasing numbers of scientists are feeling optimistic that we may find evidence that we indeed live in a multiverse. With all these new ideas, it's clear that reality indeed is likely much stranger than most of us realize.

In recent years, two-thirds of physicists surveyed agreed that, like Schrödinger's cat, you and I and everyone and everything else exist in a superposition of states. The Schrödinger's cat experiment was designed by physicist Erwin Schrödinger to express his concern that something as large as a cat could be placed in a superposition of states inside a closed quantum system. In this thought experiment, a hypothetical cat is placed inside a sealed box with a flask of poison and a quantum random triggering device, that might at any moment break the poison flask and kill the cat. The paradox is that before the cat is observed, it exists in a superposition of states—simultaneously both dead and alive.

Another huge development is that physicists have demonstrated that there may be no such thing as objective reality. Two observers can literally be in the same place at the same time, while experiencing completely different realities—with both considered to be correct, and neither more accurate than the other. This is not to say that two observers will never witness the same reality, but rather that very different observations can be made by observers positioned next to one another, and both observations can be considered to be true. [1-19]

LOOKING TO SCIENCE FOR DEFINITIVE PROOF

Our existing scientific method was founded on assumptions that don't incorporate the reality of quantum physics. Quantum weirdness such as "spooky action at a distance," quantum tunneling, quantum teleportation, quantum superposition of states, and quantum coherence are a well recognized part of the weird, wonderful world of the realm of the quantum realm of the very small—but all of these quantum properties are increasingly also acknowledged above the "quantum realm." When quantum mechanics arrived on the scene circa 1900, many scientists fervently hoped they would not need to revise the way scientific studies are conducted. They assumed scientific assumptions still hold true. Yet quantum physics consistently shows that is likely not the case.

Material realism's first assumption of strong objectivity asserts there is an objective material universe "out there" that is independent of us. For example, we hope to understand chemistry by considering the interaction of chemicals, and physics by studying how physical bodies interact. In this model, we do not consider ourselves to be in any way entangled with what we are observing. Classical scientists imagine they can be perfect voyeurs, having no appreciable impact whatsoever on the subjects of study. The validity of this assumption becomes questionable when we see how inextricably connected the observer is in quantum physics experiments, where the observer plays a decisive role in determining what is observed.

Material realism's second assumption of causal determinism is that once we understand the forces of change influencing a given system, we can accurately predict the effect those causes will have. For example, we might expect to accurately predict what would happen if we roll a marble toward two other marbles, provided we know where all the marbles are located, and the position and speed of the striking marble. Quantum physics challenges our ability to understand all that we need in order to make such predictions. The uncertainty principle of quantum physics does not allow us to know both an object's velocity and it's position at the same time; we can know one or the other, but not both. Werner Heisenberg proposed in his uncertainty principle that quanta must be described as probability waves while they travel, and as particles (matter) when they are viewed. Factoring in such quantum influences, we cannot accurately make definitive predictions, other than establishing statistical probabilities for specific outcomes.

Material realism's third assumption of locality is the concept that objects exist independently and separately from one another. In classical scientific experiments, scientists study an experimental subject with the assumption that it is separate from others, with the belief that they are studying an isolated, independent subject. This assumption breaks down in quantum physics, where it is possible to observe twin entangled quantum particles that remain in synchronization with each other across time and space, continuing to correlate their angles of spin with each other. Classical materialistic science has not accommodated the reality of such non-local connections.

Material realism's fourth major assumption asserts that subjective mental phenomena can be dismissed as simply being epiphenomena of matter. In other words, the material world is assumed to be the primary mover and shaker, and things we think we observe through our minds are merely irrelevant, inconsequential side effects that can for all intents and purposes be disregarded. Quantum physics reveals the weakness of these assumptions when we see how observers who take quantum measurements have the power to collapse quantum waves at the moment of observation.

For many years, most scientists have accepted material physicalism as the gold standard by which we can tell what is real from what is not. It's thus hardly any surprise that we feel shocked to discover that the way we remember a given brand, song lyric, logo or movie dialogue is completely different from what now seems to be true. Thanks to people sharing Mandela Effect experiences, we begin to see that some of our alternate memories may not match the official historical records, but they might match the memories of many other people. And there is strength in knowing we're not alone, and we're not necessarily confused or mis-remembering.

MANDELA EFFECT SURVEYS

By August 2022, Mandela Effect experiencers appeared to represent a majority of the population; YouGov's poll results showed 60% of one thousand people surveyed remembered the Berenstein Bears. [1-20]

RealityShifters surveys conducted in 2000 and 2013 indicated that despite small increases of reports of some types of reality shifts, the percentages of experiencers are roughly similar. In April 2000, I conducted a survey of 395 participants, with the goal of determining what types of reality shifts (aka "Mandela Effects") are most commonly observed, and how people feel when encountering these experiences. 72% of respondents in the 2000 survey said they'd witnessed things going missing, and 45% said they'd seen people, animals, and things appear that weren't there before. 51% noticed things having changed, or been transformed. [1-21]

In May 2013, I followed my 2000 survey with a second "How Do You Shift Reality?" survey—involving a statistically significant group of 567 respondents. In 2013, 69% noticed things going missing; 48% had seen people, animals, and things appear; and 44% noticed things having changed, or been transformed. An additional question was added to the 2013 survey, where 27%, or 144 of the 541 respondents, indicated that they'd seen dead people and animals alive again. [1-22]

WORLD WIDE PHENOMENON

The Mandela Effect is a truly world-wide phenomenon which has been appreciated and noted by many Mandela Effect researchers. The International Mandela Effect Conference (IMEC) features international reports of Mandela Effects such as those observed in Brazil, including differences observed in statues and the national flag.

Since 2016, Carter Tweed's website, AlternateMemories.com, has conducted surveys of Mandela Effect-affected people, focusing on the question, *"Can the past change?"* The AlternateMemories website shows top Mandela Effect categories to be: Art, General, Books, Religion, Movies, Music, Science, Geographical, Brands, TV, and people. Tweed reports that the top ten countries reporting Mandela Effects include: Canada, United States, Australia, Netherlands, United Kingdom, New Zealand, Italy, Germany, and France. [1-23] Some of the top Mandela Effects being noticed world-wide include changes to human anatomy, such as changes in the location of the human kidneys and heart. While people living in Africa seldom recall Nelson Mandela having died while incarcerated, citizens of countries outside of the African continent more often recall that this occurred.

Thousands of Mandela Effects have been documented in dozens of countries over the past twenty-five years, with evidence of the global nature of the Mandela Effect documented on both the *"your stories"* pages of the www.RealityShifters.com website, and the monthly *RealityShifters* ezine. These records show which Mandela Effects were being noticed at what time and in which countries, starting in October 1999. [1-24]

A remarkable Mandela Effect example on the *RealityShifters* website was noted in the Danish TV series, *Matador*. Denmark's state-owned television channel created this series, depicting small-town life in Korsbæk, Denmark in the years before, during, and after World War II. This popular series originally aired between 1978 and 1982, and has been subsequently rebroadcast. The Mandela Effect involves a particular scene in which an old woman, Misse, had been married to a teacher named Andersen. In this scene, the old lady, Misse, tells her friend, Hans Christian, that she has shut her husband (Andersen) out on a balcony when he tried to get up in her bed on their wedding night. Andersen had gone out onto the balcony, because he had heard a shot, and while he was locked out on the balcony, he died in the rain. The Mandela Effect involving this TV show is that most every Dane who watched the show remembers having seen Andersen, the unfortunate husband, out on the balcony wearing a nightshirt (not pajamas). And yet, Denmark's radio claims people are experiencing mass hysteria, because that scene was never filmed—and there is no missing footage. This is still a hot topic of conversation in Denmark, with *Danish Radio* writing:

> *"Most probably also remember the pictures of how the horrible teacher, Andersen, in white night shirt is locked out on the balcony on his wedding night by the wife, Misse Møghe. The fun is that the situation has never been shown on television, but only described through conversations between Misse and Hans Christian and his brother Jørgen Varnæs. However, we still remember the balcony picture so painfully."* [1-25] [1-26]

The official explanation for Danish people remembering a balcony scene where Andersen wore just his nightshirt, and was locked out on the balcony by Misse Møghe, is that people simply imagined having viewed his scene, because the other characters talked about it so vividly. The official explanation is that the balcony scene with Andersen was:

> *"... only described through conversations between Misse and Hans Christian and his brother Jørgen Varnæs."*

Yet thousands of people recall having viewed a scene that no longer exists. I haven't seen this TV show, but I can relate to witnessing changes in movies and TV shows, since I've witnessed my own copies of DVD movies changing from one viewing to the next—sometimes with an entire scene no longer included in a movie that I watched and that I own, that had continuously been in my possession.

POSITIVE MESSAGES IN MANDELA EFFECTS

Mandela Effects are viewed by some as being positive and helpful—possibly due to expecting or being open to this being true. Shane Robinson noted that Mandela Effects appear to be largely positive, innocuous and gentle—as if the cosmos is providing us with improvements in our way of life that are primarily helpful, rather than harmful. Canadian artist Kimberly-Lynn Hanson's "Closer to Truth" art installation presented in 2019 invited viewers to consider how each example of a Mandela Effect, such as Jiffy peanut butter changing to Jif, is closer to truth. In my book, *Reality Shifts*, I describe how the experience of such physical changes in reality can be viewed as living lucidly—as living in a waking dream where our thoughts become reality and we are continually engaged in a conversation with the cosmos.

I'm grateful that out of all the people noted as being alive again, Nelson Mandela is the one who's best recognized. While the very first person mentioned in a book as being "alive again" was Larry Hagman in my book, *Reality Shifts*, the "Hagman Effect" does not convey quite the same level of positive message that Mandela's legacy brings. Nelson Mandela's positive example is well aligned with the Mandela Effect, that challenges humanity to explore what it means to be *homo sapiens*— wise people—through this exceptional human experience.

INTERNATIONAL MANDELA EFFECT CONFERENCE

In March 2020, the International Mandela Effect Conference (IMEC) formed a 501(c)3 non-profit organization, "to inspire, empower, and unify all curious observers. Through the evidence presented as the Mandela or Quantum Effect we bring into this phenomenon's own alignment truths which have proven themselves timeless in all religions, spiritual practices, and scientific findings." IMEC's first event was held in Idaho in November 2019, although the official organization was yet to be formed.

The June 2020 IMEC conference, *Revelations of the Mandela Effect*, took place online during the international coronavirus pandemic. This live-streamed event featured pre-recorded talks and live panel discussions— including one by Eileen Colts, describing how she'd interviewed Nelson Mandela while working as a journalist, and then, a few years later, covered the news story of his death in 1987, before being very surprised to see him alive again in London. The IMEC 2020 conference was conducted through the International Mandela Effect Conference YouTube channel, where they can be viewed any time.

One episode of IMEC *Open Tables* focused on Mandela Effects in Brazil, with some Mandela Effects discussed on that show including: Brazil being called "Republic of the United States of Brazil" until 1967; changes to Christ the Redeemer statue, with his heart in the middle of his chest and a different, larger statue base; changes to the monument dedicated to the Bandeiras ("to the flags") nicknamed "let me push" because it used to be that people were pushing the boat; changes to the Brazilian flag, with the lone star being farther above the white strip, and the words *"Ordem e Progresso"* in black, not green; and Brazil and its continent now being closer to the African continent, with a time zone one hour ahead of the USA. [1-27]

REDEFINING THE MANDELA EFFECT

Just as there are many names for the Mandela Effect, there are many definitions of it, too. The best terms and definitions will remain descriptive without jumping to conclusions about what is causing the Mandela Effect, while providing descriptive terminology such as *mismatched memories* or *alternate recollections*, rather than such terms as "false memories" or "misremembering" which unnecessarily invalidate memories that do not match historical records. The best Mandela Effect definitions will acknowledge that this phenomenon can involve any number of people, and that thanks to this phenomenon, we are gaining a much deeper understanding of the majesty of Nature, the cosmos, and reality.

* * * * * * *

This book is structured in 11 chapters: (1) this introductory chapter, (2) an overview of the ideas, philosophy and society of those experiencing the Mandela Effect including a glossary of common terms, (3) examples of my personal Mandela Effect experiences, (4) a brief history of the Mandela Effect, (5) theories as to why the Mandela Effect occurs, (6) the science of the Mandela Effect including a glossary of relevant terms and ideas, (7) a more detailed look at what Mandela Effect experiencers have in common, (8) recognizing the Mandela Effect as an Exceptional Consciousness Experience, (9) finding gifts in experiencing the Mandela Effect, (10) consideration of how the Mandela Effect can provide what humanity needs at this time of humanity's great awakening, and (11) Questions and Answers. Practical exercises and applications are included at the end of each chapter, allowing readers to engage with the material and apply newfound knowledge.

As more people begin to recognize the existence and significance of the Mandela Effect, we may wonder what groups of Mandela Effect experiencers have in common. Are there shared norms and values? If so, how are these communicated? The next chapter, *Mandela Effect Society,* provides an overview of the society of Mandela Effect-affected individuals, with some of the terms and language unique to this community.

~~~~~~

EXERCISE:
YOUR FIRST MANDELA EFFECTS

Which Mandela Effects did you notice first? Perhaps you
subliminally noticed reality shifts such as traffic lights having
switched whether green light or red light is on top. Or you might
have had a memorable personal moment where you knew
something is different than it has "always been."

~~~~~~

EXERCISE:
YOUR BIGGEST MANDELA EFFECTS

Which Mandela Effects are biggest for you? Which are easiest for
you to acknowledge as something you know has changed? These
can be quite distinct, personal and individual, and are not proof or
evidence that some people are right—it's simply a matter of
acknowledging changes that feel most striking to you.

~~~~~~

EXERCISE:
MANDELA EFFECT GROCERY SHOPPING SAFARI

Once we start noticing the Mandela Effect, we can see it
everywhere. It can become impossible to make a trip to the grocery
store without running into items that have changed. From
unhyphenated Kit-Kat candy bars, to Haas avocados labeled
"Hass," just about every aisle provides reminders of the ubiquity of
the Mandela Effect. A fun way to investigate this is to take a
Mandela Effect shopping safari. Be alert to noticing changes in the
names of foods you thought were spelled differently, or had a
different logo. Pay attention to grocery store items you don't
necessarily purchase, but that you are familiar with. For extra fun,
take pictures of items you remember differently, so you can further
investigate. Make a checklist of items that look different than you
recall. Such foods might include: Oscar Meyer hot dogs, Haas
avocados, Fruit Loops breakfast cereal, Captain Crunch breakfast
cereal, Jiffy peanut butter, Double Stuffed Oreos, Kit-Kat candy
bars, Pixie Stix candy, Cheese-Its baked snack crackers, Stouffer's
Stove Top Stuffing, Cup o' Noodles, Coke Zero, and Febreeze.

~~~~~~

"Sometimes,
it falls upon a generation to be great.
You can be that great generation.
Let your greatness blossom."

—Nelson Mandela

Chapter 2

MANDELA EFFECT SOCIETY

How do you feel about the notion of widows eating fish? If you were raised outside of India, this idea probably doesn't trigger any strong emotions one way or another. Citizens of the Indian village in Bhubaneswar would likely feel uncomfortable about this question, since they associate fish with aphrodisiacs, and expect widows to respectfully refrain from eating such foods after their husbands have passed away. Those of us who are not offended by the idea of widows eating fish have a cultural blind spot with regard to this matter. Cultural clashes can occur where social taboos are not respected; that what is considered normal in one culture might be immoral in another. Ethnocentrism, where one presumes one's own culture to be superior to others can embolden those who feel their beliefs represent majority opinions, while simultaneously contributing to marginalization of people who are not part of that majority. [2-1]

Taboos reveal our expectations regarding normal customs, values, and traditional ways of doing things, with a social understanding that at some point, what falls far enough outside of those expected traditions is considered unacceptable. We can become so fully ensconced in our cultural assumptions that we can become blind to them, assuming that our worldview is adopted everywhere, by everyone. We can then be quite surprised to meet people who believe differently.

Experiencers of Mandela Effect share a common language, with Mandela Effect terminology and symbols corresponding to areas that Mandela Effect-affected (ME-affected) people consider important. ME-affected individuals each have unique experiences, backgrounds, and beliefs—including unique political and religious beliefs—yet come together with a sense of shared purpose about understanding the Mandela Effect phenomenon.

Author and researcher Aaron J. French describes a few distinct types of Mandela Effect experiencers, including "Christian conspiracy theorists' and "New Age occultists." [2-2] While these subgroups are part of the Mandela Effect community, there exists far greater diversity than just these smaller sub-groups in the larger community.

33

MANDELA EFFECT NORMS, VALUES, AND BELIEFS

Because there are so many differences observed in Mandela Effects, many online Mandela Effect communities set ground rules to foster respectful communication within ME groups, discouraging statements such as, "That is wrong," or "It's never been that way." Mandela Effect experiencers appreciate the courage required to step forward and inform friends, family, and coworkers that they remember things differently. Because no two Mandela Effect experiencers share exactly the same sets of memories, tolerance of different viewpoints and experiences is recommended.

Another valued characteristic of the Mandela Effect culture is community networking. Mandela Effect experiencers quickly realize that talking about this topic with the general public can lead to emotional burn-out, since few people seem to care whether reality is changing. The ME community thus provides emotional support and empathy by sharing information and theories for consideration and discussion, and helping bring people together to respectfully discuss relevant concepts.

Respectful consideration of different ideas—including seemingly impossible and fringe ideas—has been the foundation for every great paradigm shift and transformative movement in human history. Most every social movement has begun with just a few core individuals aware of a better possible reality, networking with larger groups of people to discuss notions and theories considered to be outside of the norm.

INTERNET ANONYMITY

People have historically preferred anonymity when sharing personal experiences of a transcendental nature, such as the Mandela Effect. American psychologist Charles T. Tart collected hundreds of transcendent experiences written by fellow scientists from many different fields, who approached him quietly, so as to not jeopardize their careers. While these scientists felt their transcendent experiences were significant and meaningful, they knew they could not communicate any of this "to their colleagues or friends for fear of rejection or ridicule." Tart documented transcendent experiences including: altered states of consciousness, deep feelings of connection with life or the universe, apparent paranormal or psychic communications, and various kinds of apparent transcendence of our ordinary physical selves. [2-3] Tart's transcendental experiences collection can be viewed online at The Archives of Scientists'

Transcendent Experiences (TASTE), whose online journal provides a safe place for scientists to share experiences without ostracization or endangering reputations. [2-4]

Mandela Effect experiencers can join numerous online communities of open-minded, similarly-interested people to discuss experiences and theories about what is going on. These supportive communities typically encourage sharing personal experiences with consideration for those whose beliefs or opinions differ. Many online communities provide anonymity or concealed identity, so experiencers can share experiences free from concern of being judged.

I've documented thousands of first-hand experiences with alternate histories, beginning in the late 1990s. People from all walks of life have shared their personal Mandela Effect/reality shift experiences on the realityshifters website. I've heard from doctors, lawyers, engineers, artists, business people, university professors, and high-ranking professionals employed at the British Parliament, and the Pentagon. People who contacted me with their personal experiences often felt shy about "going public" with their full names, due to long-standing social stigma about being considered flighty or crazy if friends, family, or colleagues knew of their experiences.

I document the locations in the world where these reality shifts occur, including the experiencer's first name (or pseudonym), along with their geographic location, and post these accounts by month and year as I receive them, along with the dates and times that the reported events were first observed on both the "your stories" pages of the RealityShifters.com website and the monthly ezine newsletter pages. These detailed reports provide a body of evidence that is archived online on the realityshifters.com website, providing documentation of the Mandela Effect happening worldwide, for decades. These reports detail descriptions of reality shifts and Mandela Effects being noticed by country, month, and year reported, starting in October 1999. [2-5]

MANDELA EFFECT ONLINE ARCHIVES

Three notable long-standing internet archives exist of thousands of first-hand reports of Mandela Effects and reality shifts:

Since October 1999, the www.realityshifters.com site has been providing people with a forum for posting and reading Mandela Effect experiences every month.

Starting in 2010, Fiona Broome's www.mandelaeffect.com website provided people the opportunity to post first-hand accounts of their Mandela Effect experiences.

Reddit has a number of threads dedicated to the Mandela Effect, including: MandelaEffect, Retconned, Mandela_Effect, Glitch_In_The_Matrix, and GlitchInTheMatrix.

MANDELA EFFECT YOUTUBE COMMUNITY

YouTube Mandela Effect communities allow opportunities for Mandela Effect-affected people to join together in live chat comment areas on livestream feeds. Mandela Effect experiencers share changes they witnessed in specific areas, such as: geography, physiology, historical events, products and logos, movie dialogue, animals returning from extinction, and more. Many YouTube channels host live broadcast events providing regular gathering places on YouTube channels such as: International Mandela Effect Conference (IMEC), the Quantum Businessman, Moneybags73, Shane Unbiased and On the Fence, and Once Upon a Timeline. Other noteworthy YouTube channels covering the Mandela Effect include: Open Your Reality, Guy Fauqes, Vannessa VA, Brian MacFarlane, Brian Stavely, and my *Cynthia Sue Larson* YouTube channel, covering Mandela Effect-related material since 2009.

YouTuber MoneyBags73, aka Evan Matraia, created a series of hundreds of Mandela Effect "voting videos," which provide running totals showing numbers of people who recall a given Mandela Effect (voting with a "thumbs up") and those who do not (voting with a "thumbs down").

GEOGRAPHIC AND COSMIC CHANGES

One of the most amazing categories of Mandela Effects involves people remembering geography differently. While these changes can be recognized in familiar, local areas, they are most often observed on maps or charts describing places farther from home. Similarly, geographic changes are less often noticed by subject matter experts, perhaps due to the Quantum Zeno Effect (QZE). The Quantum Zeno Effect operates much like the old adage, "a watched pot never boils," which suggests that the more we stare at something or frequently check or observe it, the less likely we are to observe any change. We might then expect that medical experts won't be most likely to witness Mandela Effect changes to human physiology, and South Africans might not have ever thought Nelson Mandela died while he was imprisoned. So when it comes to geographic changes, we seem most surprised to witness changes outside of our immediate area.

The absence of a North Pole continent has surprised many people. I recall having seen "North Pole" featured in seasonal holiday cards and books that sometimes included signs for Santa's workshop along with a red and white striped "North Pole." I remember looking for the North Pole on globes when I was a little girl, and feeling delighted to find the place where Santa lived. Much to our surprise, maps and globes now show a large open expanse of water at Earth's North Pole, with no ice, snow, or possible way to post any such signs whatsoever.

Many people notice that the continent of South America seems to now be much further "off to the right," or pulled in an eastern direction from where we recall it used to be. Not only do maps seem quite different from what we recall, where the northern and southern continents were more closely aligned, but some people also recall a time when time zones used to be more closely aligned between people living in, for example, the USA and Brazil. Brazilian author Carla Cicarino discussed how now, in South America, "we are almost kissing Africa," and how the continents used to be much more vertically aligned, such that it was almost a straight line up from San Paola to New York, sharing a time zone—whereas now, Brazil is in a time zone that is one hour ahead of the USA. Carla asked friends to draw the continents the way they remember from school, and saw people draw maps the way she remembers. Investigations to see if there were any ancient maps drawn the way people recalled didn't turn up any clues as to why people were so convinced that the map is so different. [2-6]

Sardinia and Corsica seem much larger than many people remember, and Sicily seems now to be much closer to Italy. [2-6]

Some people who remembered a time when Gibraltar was known as the "Isle of Gibraltar" were shocked to see maps showing the Strait of Gibraltar, with Gibraltar now consisting of a peninsula that is attached to southern Spain. For those remembering the "Straits of Gibraltar" (plural of strait), that name made sense, because they remembered a northern outbound strait and a southern inbound strait, with what had been known as the "Isle of Gibraltar" being an island in-between.

Some people notice that the sun seems brighter, more intense, and whiter now than what they remember in years past. For those noticing this change, the sun is often recalled as having seemed to be more yellow, and less harsh.

Perhaps the biggest of all geographic changes involves our location in the Milky Way galaxy. As Carl Sagan once said on his TV show, *Cosmos*,

> *"... is another similar galaxy, its spiral arms slowly turning once every quarter billion years. This is our own Milky Way, seen from the outside. This is the home galaxy of the human species in the obscure backwaters of the Carina Cygnus spiral arm."*

Curiously, the location that Sagan so carefully describes no longer fits our current location in the Milky Way. *Universe Today* places us about halfway between the center and the edge of the large, flattened disc that is the Milky Way. Previously, we'd been situated out at the edge. Carl Sagan had made our fringe location abundantly clear when he stated, *"The fact that we live in the outskirts of the galaxy,"* while pointing, *"way out here,"* at the outer edge of the Sagittarius arm. [2-7] [2-8]

PHYSIOLOGICAL CHANGES

Of all of the various types of Mandela Effects, changes to our physical bodies often hit closest to home. Numerous changes to human physiology have been noted, affecting organs and the skeleton, and sometimes appearing to be upgrades or improvements. Examples of such improvements can be seen in changes to the kidneys and heart.

Those employed as doctors, physicians assistants, and nurses are subject matter experts who tend to more frequently observe human anatomy and physiology, which has the effect of "locking in place" established facts and histories in their areas of expertise. Therefore, they may not be expected to notice MEs in these areas.

KIDNEYS

The official historical location for kidneys is no longer in the lower back, but instead higher up, closer to the lower rib cage, so they are now more accessible to surgeons from the front of the body. The name of the so-called "kidney punch" in martial arts no longer makes sense, since that target location is no longer where the kidneys are situated. Comparing descriptions of "kidney punch" and kidney location, we see why so many people are amazed by this particular shift. Here's Google's description of our new-and-improved kidney location:

> *"The kidneys are bean-shaped organs (about 11 cm x 7 cm x 3 cm) that are located against the back muscles in the upper abdominal area. They sit opposite*

each other on both the left and right side of the body; the right kidney, however, sits a little lower than the left to accommodate the size of the liver."

Google's description of "Kidney punch" shows:

"A kidney punch is a punch that occurs usually when the fighters clinch. It is a hit that goes into the lower back, to the kidney area. It is illegal because of its high danger level to health."

I clearly recall that kidney punches used to be considered potentially lethal. I remember hearing that such strikes would occasionally result in the recipient urinating blood–a very bad sign of kidney damage–which apparently no longer represents such a danger. A publication of the U.S. Navy from 1914 specifically refers to how one Navy boxer took advantage of the Navy's *"no kidney punch"* rule:

"Sally made the most of the 'no kidney punch,' and whenever the going got heavy he would rush into a clinch and turn his exposed back and kidneys up, a la roll top desk. The sailor would raise his glove and then change his mind about smashing it down on Salvadore; he wanted to swat Sally's back so bad he could taste it!" [2-9]

One woman was astonished when I mentioned during a talk I gave in San Jose in August 2017 that there had been a change in the location of the kidneys. She had noticed something unusual at a recent visit to her doctor for her routine annual kidney exam. Much to her surprise, the doctor accessed her kidneys from the front, near her lower ribs. She asked, *"What are you doing?"* since this was completely different from the way she'd been examined in the past. When the doctor replied, *"What do you mean?"* she continued, *"You always have me lie down on my stomach, so you can reach my kidneys."* The doctor said the procedure was always done from the front. She was amazed that this change had just come to her attention the same week that I included it in my talk. From the doctors' perspective, the kidneys have "always been accessed from the front," yet she definitely remembered something quite different.

HEART

The human heart is now located close to the center of human chests, rather than slightly to one side. Many people remember being told by teachers at school that the heart was slightly to the left. I remember placing my right hand over my heart to say the pledge of allegiance each morning in elementary school. Our teacher would ask us to, "Place your right hand over your heart," sometimes taking advantage of a learning opportunity to point out that this position was slightly left of

39

center, for anatomical accuracy. Some of us also remember that when getting our blood pressure taken, our left arm was proffered, since it was common knowledge that the left arm provided the best way to get a clear reading, because it was closer to the heart.

Many more physiological changes have been noted, including: spine, hand, and skull changes, and body temperature changes.

CHANGES TO HISTORICAL EVENTS

Many famous historical events are remembered quite differently than stated in current historical records. People recall having seen the Zapruder film footage that clearly showed there were only four people seated in two rows of seats in the president's car on November 22, 1963, when president John F. Kennedy was shot in Dallas, Texas. Abraham Zapruder's film now shows there were six people in the car, and some people say they witnessed flip-flops on this Mandela Effect, going from four to six and back to four people again.

The "Black Tom explosions" occurred prior to America's involvement in World War I, on July 30, 1916 in Jersey City in the New York Harbor. Many people do not remember this famous event that is one of the largest artificial non-nuclear explosions to have ever occurred, and the first large-scale terrorist attack on US soil. The explosions killed four people, destroyed some $20 million of military munitions, and damaged the Statue of Liberty. This attack by German agents was so large that people as far away as Maryland were awakened by what they assumed was an earthquake. What's fascinating about this event is that many people, including some employed by the US military who attended military historical training, never heard about it. Many Mandela Effect experiencers feel certain they would not have forgotten about such a pivotal historical event, yet it is new to them.

Other examples of changes to historical events include such things as: "Tank Man," the "Hooligan Navy" of volunteers during World War II, changes to Easter Island Moai megalith statues, and a change involving a "new" Hurricane Erin now having been set to make landfall in New York on the same day as the attacks on September 11, 2001.

MEETING PEOPLE WHO WEREN'T PREVIOUSLY THERE

I remember hearing the news of Nelson Mandela's death when he was incarcerated many years ago, and again more recently. I also recall having heard news through the international news media of Jane Goodall's passing away within a month or two after Dian Fossey was killed in December 1985, which struck me as such a senseless double loss of great female primate researchers in mysterious rapid succession. I was thrilled to learn that Jane Goodall is still alive and well, and grateful to see her give a talk in April 2014 in Berkeley.

If people who died can be seen alive again, can a person who didn't previously exist, suddenly appear? There is a satisfying logic to this, in the sense that just as one might find a lost item suddenly appear as if out of nowhere in an unexpected place, it's equally possible to find a new person suddenly appear in our lives when we need them most.

In 2015, two people informed me that I had come into their lives as if out of nowhere. Both of these people separately told me, "If you had existed, doing what you're doing, I would have known." When conversing with me about this possibility that I hadn't existed in previous realities, one person exclaimed, *"It may interest you to know that apparently you didn't exist (at least not in your current version) in the timeline I came from, because if you had, I would have heard of you a LONG TIME AGO!"* I replied, *"How fascinating that you hadn't come across me, so I'd tend to agree with you that in previous realities you've come from, I didn't exist as I do here. Cool!"* A second person told me she'd found me in January 2015, when she suddenly realized we'd been friends on Facebook for quite some time, and had mutual friends. She felt certain that I'd popped up out of the blue, as it were, since she was positive she'd have contacted me much sooner. These people both told me that they were certain they had been in realities in which I had not written books, published a newsletter, or conducted research in the field of reality shifts, quantum jumps, and consciousness.

I've noticed some authors seem to pop up out of nowhere. A couple of noteworthy examples for me include Neville Goddard and Dolores Cannon. In both cases, I was reading extensively in the topics they specialized in during years when I would definitely have come across their books in libraries and bookstores, yet I did not hear about either of them until many years later.

I've had a remarkable experience with UC Berkeley's Dr. Yasunori Nomura, who I first came across many months after having written my book, *Quantum Jumps,* in 2013. I had not seen any of his papers, despite conducting a thorough search of research papers written in his field of expertise. Nomura is now shown to be the first to have published ideas about the many worlds of quantum physics being one and the same as the multiverse. I appreciate his papers for their clarity of presenting

41

complex material, and I definitely would have included references to his research when writing my book, *Quantum Jumps*, if it had been there for me back in 2013. I'm deeply grateful that Nomura apparently quantum jumped into my reality! [2-10]

We can ask ourselves, *"How good can life get when people show up out of the blue?"* and enjoy living the answer to that question. Finding a mentor, mate, friend, colleague, or partner when we need them most can be one of the greatest blessings we could ever possibly receive.

TELEPORTED TO SAFETY

A surprisingly common, yet seldom discussed type of personal Mandela Effect is: being teleported to safety. Dozens of people have approached me after I have have given talks about reality shifts, to privately share that they have been teleported to safety at just the right moment when facing impending danger—sometimes arriving many yards away from where they'd just been, with no obvious means by which they could have moved. Numerous first-hand reports of this phenomenon have been included on the "your stories" and monthly news pages of the RealityShifters website.

My book, *Quantum Jumps*, shares reports of a couple of young boys who were frequently observed teleporting across many miles, and also a famous experience involving teleportation across the Pacific Ocean. Teleportation seems to happen most often to individuals whose lives are endangered by something happening—such that the teleportation saves their lives.

INTUITIVE NONLINEAR LIVING

The sundials showing up out of nowhere that Christopher Anatra and I separately saw in our personal Mandela Effects feel to me like an invitation from the cosmos to return to natural time—to let go of human timepieces and embrace an intuitive, nonlinear relationship with time. We can return to a natural way of living in harmony with natural rhythms and cycles—and in so doing, restore a sense of meaning, purpose, and connection in our lives.

As Glenn Aparicio Parry writes in his book, *Original Thinking*, indigenous people have a unique view of our relationship to the cosmos, including the meaning of concepts such as "original" ideas. Indigenous wisdom views all ideas as being originally present in the cosmos, rather than there being "new" ideas springing forth.

Indigenous peoples have a non-linear, cyclical conceptualization of time. [2-11]

Somewhere over the course of the past several hundred years, intuitive direct knowing took a back seat to the left-brain, rational analytical focus on what was materially measurable. Many modern civilizations lost respect for subjective direct knowing based on dreams, visions, and imagination—although the majority of inventions and scientific discoveries still secretly arose from this realm. [2-12]

Organizers of the 2019 Idaho Mandela Effect conference experienced a sense of intuitive nonlinear living, evidenced by seamless cooperation and participation, and a palpable feeling that the event seemed to emerge from its own sentience. Despite the speakers not coordinating talks with one another in advance, a remarkable flow from one talk to the next occurred. Several of us were aware of moving out of "left brain" thinking to a more balanced and inclusive intuitive "right brain" way of making choices, featuring unpredictable, seemingly random aspects.

Kimberly-Lynn Hanson described the process of living a "nonlinear lifestyle," describing a game that she teaches people—including young children—ways to practice the art of discovering the sheer fun of following the lead of whatever seems to be totally random, and allowing oneself to be guided by that—the more thoroughly random, the better. Hanson encourages us to, "Follow that random!" From such a solidly random starting point, we can proceed forward in whatever direction we feel inspired to go, with some kind of general goal in mind, such as looking for "Wisher Washers," the little washers one sometimes sees on the ground from time to time, or hearts.

While working on writing this book, I consulted John Gribbin's book, *Deep Simplicity*, that I'd checked out from the library earlier before attending the conference. I'd read up to about page 32, when I felt a strong intuitive nudge to skip forward and read what's on page 94. So I skipped forward to page 94, which had a nice graphic depicting something called a Sierpinski gasket. I got a clear sense that I should read this section particularly carefully, which I did. I was stunned to see Gribbin stating that one can play a game in creating Sierpinski gaskets, which bear a remarkable resemblance to the Wisher Washers I'd just heard Hanson talking about. Gribbin emphasizes that when playing such a "chaos game" in which one produces a Sierpinski gasket from simple iterative rules, there are two important things to keep in mind:

"The first is patience—it will take several hundred steps before you really see anything like the Sierpinski gasket. The second is randomness; don't be tempted to use a computer, unless you are a good enough programmer to be sure that you are using genuinely random numbers." [2-13]

At this point, you could knock me over with a feather! Here I was reading in John Gribbin's excellent description of what he called a "chaos game" to create something that looked like a triangular gasket (with a triangular hole in the center), and Gribbin was emphasizing the importance of being thoroughly and properly random, and patient. This was astonishingly similar to what Kimberly-Lynn Hanson had suggested just a couple of weeks before!

Intuitive nonlinear living shows us that we can benefit by adopting an attitude of patient persistence, trusting that we can follow randomness. This suggests there exists a hidden order of benevolent, co-creative, conscious agency that we can play with, to wonderfully positive effect.

READING SIGNS AND SYMBOLS

"We never change things by fighting the existing reality. To change something, build a new model that makes the existing model obsolete."
—Buckminster Fuller

Are Mandela Effects trying to tell us something? If so, how can we hear their message? Many Mandela Effect experiencers seek to find meaning in what is going on with the changes being observed. When we look more deeply, we can find magic, miracles and meaning in Mandela Effects. These messages can be gleaned by reading symbolic events as dream symbols, starting from the philosophical viewpoint that the cosmos fundamentally arises from consciousness, as many spiritual wisdom teachings suggest. We can then imagine we are engaging in constant communication with All That Is. This conversation happens via our presence—through our thoughts and feelings—such that we are constantly receiving answers to questions that we often don't even consciously realize we've asked. As we focus attention on how our thoughts and feelings are asking Nature questions, and in how we are receiving answers in the form of events unfolding in our lives, we can bring this dialogue into conscious awareness.

A great way to start reading signs and symbols is with a practice of documenting and analyzing your dreams. This practice develops a mindset by which you can eventually interpret real-time symbolic conversation with the cosmos. When we view reality as being like a dream, we can appreciate that from time to time, material things and events spring forth from consciousness. We can thus find meaning in Mandela Effects by doing a four-step dream analysis approach:

(1) Write a narrative describing what happened, including all of the important details;

(2) With a fresh piece of paper, review the narrative from step one, and pick out the key words/ideas (nouns and verbs);

(3) Looking at each key word/idea from step two, free-associate and write down every idea and word and phrase that comes to mind—especially the first things that come to mind; and

(4) Taking only the free associations that you wrote in step three, create a new narrative that provides deep meaning and reveals hidden insights.

Pieces of the art installation that Kimberly-Lynn Hanson created for our 2019 Idaho conference twice fell to the ground during the event. At the time of each occurrence, which happened days apart, the art fell immediately after a speaker or panelist discussed the idea of questioning authority. Kimberly was out of the room the first time it happened, so I texted her the news that her art work had fallen, and might need some repair. Kimberly asked, *"Was it Darth Vader!"* And I replied, "Yes," so Kimberly replied, *"Ha ha ha. Of course it was :)"*

Such meaningful coincidence is noted by many Mandela Effect experiencers, who appreciate receiving meaningful two-way conversation with higher levels of conscious agency.

MANDELA EFFECT AS PARADIGM SHIFT

New experiencers of the Mandela Effect often feel a sense of culture shock, from the intensity and confusion implicit in witnessing alternate histories, while questioning their memories. Some Mandela Effect experiencers recognize that new forms of plants and animals seem to be springing up—some of them being animals long considered to have been endangered or extinct, and others previously only imaginary. The experience of the Mandela Effect can thus feel quite surreal, challenging us to open our minds to consider new possibilities, including some that may seem improbable or impossible.

With all of this inquiry, the Mandela Effect community challenges current assumptions while questioning agreed-upon facts. American physicist and philosopher Thomas Samuel Kuhn coined the term, "paradigm shift," in his book, *The Structure of Scientific Revolutions,* to describe how scientific progress occurs through a cycle of scientific models experiencing drift, crisis, and revolution that results in an entirely new understanding of the world. Such transformations often require a generation to accomplish, and also a degree of faith, as well as a good degree of reason, for such wide-sweeping cultural shifts to be complete. [2-14]

HIDDEN SYMBOLIC MEANING OF THE MANDELA EFFECT

I appreciate how the Mandela Effect was named after South African president Nelson Mandela, because his inspirational and wise words are perfect for expressing key ideas that are relevant to the Mandela Effect. When viewing the cosmos through nonlinear eyes, we begin to see how the name "Mandela Effect" beautifully emphasizes an underlying principle of the Mandela Effect. A subliminal focus on freedom imbues the Mandela Effect with brilliant strength of spirit, in a call for humanity to join together and embrace the pure indomitable spirit that is a birthright for each and every one of us.

It's been said that if you want to understand a culture, you need to understand it's language and terminology. Linguist Benjamin Lee Whorf proposed that language profoundly influences the way we think. We can thus seek insights into the mindset of the Mandela Effect and its community through language that has specific meaning to those experiencing this phenomenon. The following section, "Mandela Effect Terminology" is a list of key words and phrases associated with the Mandela Effect, together with short definitions and descriptions for each term. Not all of these words are utilized by all Mandela Effect experiencers, and terms and language are ever-evolving. This list is not intended to comprehensively document all Mandela Effect terminology, but rather provide a foundational basis for improved understanding of Mandela Effect mindset, concepts, and ideas.

MANDELA EFFECT TERMINOLOGY

Affected are Mandela Effect experiencers who acknowledge that their memories are different than recorded historical facts. Variations of this moniker adopted by experiencers include: Affected, Effected, Mandela Effected, Mandela Effect-affected, or Mandela Affected.

Akashic Records involve the idea that there exist compendiums of all events, thoughts, feelings, and ideas throughout time and space that are relevant to humanity, and that can be intuitively accessed.

Alive Again people are sometimes seen to be reported dead through reputable sources, such as major news networks, newspapers, or magazines—only to later be noticed very much alive again at some future point in time—often with no remaining trace of those news accounts originally witnessed. Nelson Mandela is an example of this phenomena. This is more often observed for people (or animals) at the periphery of our awareness, such as celebrities we hear about occasionally, rather than those we see everyday.

Alternative Recollections is an unbiased descriptive term for the Mandela Effect. (see: Mismatched memories)

Anchor Memories can serve as markers in relationship to other associated events, helping confirm that reality shifts have occurred. An example includes remembering having contemplated why specific wording was chosen for side-view mirrors on cars, *"Objects in mirror may be closer than they appear,"* which some of us recall once stated, Some people remember having seen these words in the movie *Jurassic Park*, where a dinosaur memorably appears in the side view mirror.

Awakening Experiences are unique to each individual, and mark pivotal turning points of great influence for a person's future viewpoints, beliefs, perspectives, and decisions.

Bilocation is also known as multi-location; it is the experience of someone witnessed being tangibly present in more than one physical location at the same time.

Collective Consciousness involves a coherent, agreed-upon factual history of events that is accepted by the majority as true. Local "bubbles" of remembered histories can be collectively considered to be true by some communities. (see: Downloads, Reality Bubbles, Smith Effect, and Take the Update)

Debunk is to investigate possible Mandela Effects in order to determine whether or not the perceived alternate history is real, or whether it will be disregarded. Debunkers will do well to note that just because one person did not experience a particular Mandela Effect does not mean a shift was not observed by others.

Déjà vu is a sensation where an experiencer recognizes that events currently unfolding have already been experienced at some previous time, as in a dream. Such experiences demonstrate for experiencers that we have nonlinear access to events in reality.

Dimensional Shift is a term for all types of both everyday and extraordinary personal Mandela Effect experiences. (see: reality shifts, JOTTs, jottles, DOP)

Disappearing Object Phenomenon (DOP) is the term coined by Tony Jinks for personal Mandela Effects where it seems objects appear, disappear, transform, and transport. (see: JOTT, jottles, dimensional shift, reality shifts)

Do-Overs are the idea that a sequence of events can be revisited, by going back in time in some manner—such as through hypnosis, shamanic journeying, or by utilizing some kind of quantum eraser—to make different choices than one remembers having made before, as a revision. (see: Revision Technique)

Double Memories occur when remembering things "both ways"—such as stop light colors having red on top or green on top. When more than one memory exists, it can be difficult, or even impossible, to feel certain that any one of them is "the true memory," as there may seem to be equal validity to more than one. (see: Dual Memories)

Download, as in, "he got the download," suggests that even though someone remembered something the way many other ME-affected people do, and even came up with that memory of the way things used to be, at some point the person changes their answer to match current historical facts. This is similar to "The Smith Effect," or "taking the update." Downloads can affect various levels or "bubbles" of collective consciousness. (see: Collective Consciousness, Reality Bubbles, Smith Effect, and Take the Update)

Dream Mandela Effect experiencers often sense a dream-like quality to reality, where events occur with dreamlike logic, and reality feels like a dream. Dreams provide evidence that perceptions are produced by the brain, rather than being part of the external world, and can be experienced at times as "realer than real," providing profound insights, premonitions, and information across great separations of space and time.

Awareness that all of what we presume to be material and real is actually a dream is similar to Simulation Theory. (see: Simulation Theory)

Dual Memories are remembering something in more than one way. Dual memories might involve having both a clear memory of the way something was before it changed, as well as a memory of how it was after it changed. In some cases, there may be more than two such vivid memories of a given thing, such as the exact position of the hand and arm of Rodin's *The Thinker* statue. (see: Double Memories)

Exceptional Human Experiences (EHE) is a term describing experiences that are spontaneous and transcendent, involving a new awareness of self, experiences of connection, and direct participation in reality. The Mandela Effect can be considered to be an Exceptional Human Experience. EHE experiencers mention feeling sensations such as tingling, swooning, rapid heartbeat, raised hairs, goosebumps, vertigo, and breathlessness when having exceptional experiences, as described in 1999 by Rhea A. White.

Fear Consciousness can be overcome when adopting a higher dimensional viewpoint. From a higher perspective, we gain awareness that fear is unnecessary, and is not the basis for who we truly are.

Fifth World is the anticipated next world anticipated by several indigenous people, including the Hopi. The emergence of the fifth world is expected to follow an end to the current fourth world. (see: Golden Age)

Fight the Download is something some Mandela Effect experiencers do when noticing double memories, and feeling an influence of collective consciousness and a download coming in —while also recognizing the importance of holding onto original memories, and thus, "fighting the download."

Flashbulb Memories are memories involving specific details of where we were, who we were with, what we were doing and our surroundings at the moment when a significant event occurred (such as 9/11).

Flip-Flops occur when a Mandela Effect goes back and forth between two or more states. Examples of common Mandela Effect flip-flops include: Flintstones / Flinstones, and changes to position of the hand on Rodin's *The Thinker* statue. Some consider flip-flops to possibly indicate tug-of-wars between collectives vying for different outcomes.

Free Will is the ability to independently discern, ask questions, and make choices that are not predestined nor predetermined. Chief among these choices is selection of one's perspective and viewpoint, as these greatly influence what one subsequently observes. (see: Sovereignty)

Future Memories were first described by PMH Atwater in a book by the same name, as memories of future events that can feel every bit as real and familiar as memories of past events. We can sometimes discern the difference between past and future memories by applying logic. For example, when shopping for shirts and suddenly remembering these shirts were my husband's favorites that he wore all the time, I realized that this memory was technically impossible since this was my first time seeing these shirts—at which point I recognized this memory as a future memory. (see: Predictive Programming)

Glitch in the Matrix is an idea taken from a scene in the science fiction film, *The Matrix*, where the main character, Neo, sees a cat walk past a doorway, twice. "Glitch in the Matrix" typically refers to a personal Mandela Effect experience that is striking and difficult (if not impossible) for the experiencer to ignore, and may or may not be a time loop, or some kind of time shift. (see: Looping, Resets, and Time Loops)

Golden Age is anticipated to be a epoch of time ruled by highest values including: love, kindness, and wisdom. A legend of Shambhala suggests this time of enlightenment will usher in the greatest Age of all times, with great saints and sages of the

past returning to teach true wisdom of the Ages.
(see: Fifth World)

Great Awakening is awareness of being able to shift perspective of consciousness at nodal points from fear, grief, and anger to love for others and oneself, while holistically connecting with Unity Consciousness at highest levels of conscious agency.

Hear What You Want to Hear is a phenomenon where people notice it's possible to listen to lyrics in a song and hear either one set of lyrics or another, such as the lyrics in the song, "California Dreaming," where, for example, we might either hear "prepare to pray" or "pretend to pray."

Iceberg is a chart created to sort out facts and contemplate theories, starting from most commonly agreed-upon topics near the "tip of the iceberg", and moving down toward lesser known aspects down in "the abyss."

Just One of Those Things (JOTT) refers to Mary Rose Barrington's name for seemingly innocuous occurrences of fly aways, turn ups, windfalls, comebacks, walkabouts, and trade-ins, where objects appear, disappear, transform and transport. (see: DOP, reality shifts, dimensional shifts, and personal Mandela Effects)

Jump Points are nodal points where "timelines" converge and possible alternate realities coincide, with the idea that from such points of convergence, anything imaginable can be experienced by tuning one's thoughts, emotions, and energies to match a desired physical reality. (see: Merging Timelines and Jump Points)

Karma is the idea that the sum of all of one's past choices and actions from this life and previous lives exert influence on subsequent events. Karmic awareness is advised whenever doing "timeline" or quantum jumps, such that consideration is given to what is genuinely best for all concerned.

Law of Attraction is an idea that attitudes and energies attract similar results, such that you will experience whatever you focus attention and energy on. Positive thoughts attract positive results.

Lazarus Species are species of plants and animals that are rediscovered after having been previously presumed to have gone extinct, sometimes for a decade, and sometimes for thousands or millions of years, such as the Coelacanths.

Locals is the name given to people who do not remember any of the alternate histories that are being discussed; they are sometimes referred to as "locals" of their particular reality.

Looping refers to a situation in which an observer witnesses a repeating cycle of events. Often, those who are observed to be repeating a given cycle of events will not be aware this is happening—the looping is only noticed by those observing outside of the loop. It's possible that cycles of events can repeat until awareness occurs, and a person recognizes that different choices and decisions can be made, like in the movie, *Groundhog Day*. (see: Glitch in the Matrix, Resets, and Time Loops)

Lucid Dreams are dreams in which the dreamer is awake, and able to interact with and influence aspects of the dream as it unfolds. Lucid dreaming can feel similar to lucid living, when a person feels at ease with the way reality shifts in response to one's thoughts and feelings.

Mandanimals or Mandela Effect animals, are seemingly new or transformed species of animals. Some mandanimals come into existence after first being joked about, such as the hypothetical, imaginary part lion, part tiger "Liger" drawn in the movie *Napoleon Dynamite* that now has always been an actual animal. Additional examples of mandanimals include: Cardinalfish (which shoots light out of its mouth), and the Shoebill Stork.

Mandela Effect is a name for the experience of collective mismatched memories or alternative recollections, where people share similar memories differing from official facts and recorded history. This term apparently originated via late night talk show host, Art Bell, in 2001, and was popularized by author Fiona Broome in 2009.

Mandela Effect Matching refers to the matching of experienced realities, where Mandela Effect experiencers recognize memories that differ from official history, while corresponding to similar memories that other people share.

Manifesting is an intentional mind-matter interaction process, based on the idea that desired realities are created by us via any of a number of different methods or techniques, including:

prayer, meditation, visualization, and affirmations. (see: Timeline Jumping, and Quantum Jumping).

Matrix is the idea adopted from the movie by the same name, "The Matrix," that physical reality is being fabricated and created by a deeper level of reality, and so can be manipulated, programmed, and controlled. The Matrix concept is acknowledged in many Simulation Theory viewpoints and theories.

Mayan Calendar is an ancient calendar system whose "long count" on its Great Cycle came to an end on December 21, 2012, marking the beginning of the next Great Cycle. (see: Ninth Wave)

Memory Fuzz or Fuzzy Memory has to do with the way uncertainty can creep into memories. Sometimes, this kind of fuzzy memory overlays what had previously seemed fairly certain, once Mandela Effect experiencers become aware of currently recorded history.

Merging Timelines where nodal points or jump points exist can provide access for "harvesting resources of other realities" and offer a sense of how, for example, extinct animals sometimes pop up into our reality, to the point that some biologists note that sometimes all that's needed to bring extinct species back is to return their ecosystem to good health. Canadian theoretical physicist and entrepreneur Geordie Rose of D-Wave computers once remarked that new quantum computers were being designed to "tap into parallel universes and do computations," which suggests something like merging timelines. (see: Jump Points and Nodal Points)

Message Theory is the idea that Mandela Effects contain potential messages, whose meaning can be discerned through interpretation.

Mismatched Memories is an unbiased descriptive term for the Mandela Effect. (see: Alternative recollections)

Morphing is when people observe people or things transforming and changing—sometimes dramatically.

Morphogenic Fields are informational fields that serve to organize emergence and coordination of natural physical systems (such as atoms, molecules, cells, tissues, organs, etc.) at all levels of complexity, thanks to intrinsic consciousness and inherent memory. Morphogenic fields were popularized by English biologist and author Rupert Sheldrake.

Multidimensional perspectives are inclusive of perceptions and insights beyond what can be physically measured, providing the ability to witness changes to reality that might otherwise go unobserved. This is like seeing a birds-eye view above two-dimensional Flatland with the added insight of seeing through walls, and having extraordinary powers.

Multiverse is the idea that our world is one of many parallel possible worlds co-existing side-by-side. There are many varieties of multiverse theories, many of which are seriously considered by the world's top scientists. The word "multiverse" was coined by American psychologist William James, though he used the term a bit differently, to indicate a reason to put one's faith in God, since *"Visible nature is all plasticity and indifference, a multiverse, as one might call it, and not a universe."* When we consider that the many worlds of quantum physics might be one and the same as the multiverse, we gain a sense of how Nature might literally show us doorways into other possible worlds.

Ninth Wave was proposed by physicist Carl Calleman as a wave of consciousness arriving into human awareness in 2011, featuring a short time period of 36 days, and resonating with Unity consciousness. (see: Mayan Calendar)

Nodal Points or jump points are part of the conceptualization of *timelines*, where quantum jumps occur at intersections where awareness of numerous timelines converge. For example, someone might jump to an adjacent reality where instead of catching a cold, they are just getting over it, and feeling healthy. (see: Merging Timelines and Jump Points)

Non-duality is similar to Unity consciousness; it is an awareness of everyone and everything being fundamentally connected and quantum entangled with everyone and everything else. (see: Unity Consciousness)

Nonlinear Time arises from nonlinear perspective, which entails freedom from linear thinking and linear time, such that multidimensional awareness is possible. Nonlinear time includes sensations of a timeless feeling of "no time," which can be attained through meditation, and assisted by living in accordance with natural cycles (turning off clocks, timers, and devices, in order to experience this present moment now). (see: Orthogonal Time, Timelines)

Non-player Character (NPC) is a concept introduced from video games, where NPCs represent characters that are not directed by actual human players. Also called a "bot," the idea of an NPC is like an "extra" or "background player," indicating someone with little to no observable free will or actualized individual autonomy. This concept is associated with the Simulation theory interpretation of the Mandela Effect. (see: Role Playing Game (RPG), Simulation Theory)

Now is a deceptively simple yet elusive concept of a single, present moment of subjective time. (see: Present Moment Awareness)

Oracle Indicators indicate a way of looking at regularly changing Mandela Effect examples (such as a world globe or Leonardo da Vinci's portrait of Mona Lisa) to detect changes.

Orthogonal Time was proposed by author Philip K. Dick, as a way to describe how we may move between various possible "timelines" that do not necessarily directly relate to the "timeline" we are currently on, but rather exist as if skewed 90 degrees to one side. This idea bears similarities to John William Dunne's ideas expressed in his 1927 book, *An Experiment with Time*. (see: Nonlinear Time, Timelines)

Personal Mandela Effects are Mandela Effects that are unique to and observed by one particular person, and not widespread, involving memories that differ from current facts and recorded history. (see: Dimensional Shift, DOP, Reality Shifts, Just One of Those Things, JOTT, or Jottles)

Precognition is awareness of some aspects of future events. Precognition might be realized in daydream or dreaming states of mind, and can be the starting point of a déjà vu experience.

Predictive Programming is the name created by author Alan Watts to describe harnessing the power of suggestion via the media and fiction to create a desired outcome. The premise behind predictive programming is that subtle psychological conditioning through plot and story lines can acquaint the public with planned societal changes that world leaders intend to implement later on. If and when such changes are put into effect, the public is thus already familiarized with and accepting of such changes as being expected, natural progressions, thereby lessening the possibility of vigorous public resistance or protest. Examples of evidence of future events foreshadowed in the past include several scenes in the TV show *The Simpsons,* such as one episode predicting that Donald Trump would be president of the USA. While some theorists suggest such "Easter egg" references set the stage for people to passively accept possible events, others recognize a potential for people to see stories as potential propaganda, and feel empowered to actively exert influence as conscious agents. (see: Future Memory)

Present Moment Awareness spotlights the idea of this present moment being the key moment in time to focus attention on. Through meditation, we can become aware of a sense that each moment connects with eternity. (see: Now)

Quantum Immortality is an idea taken from Hugh Everett's Many-Worlds Interpretation of quantum mechanics, implying that a conscious being cannot cease to be, thanks to the constant branching of the multiverse, and depending upon many copies of the observer. This idea and the associated quantum suicide thought experiment were first proposed by Euan J. Squires in 1986. [2-15]

Quantum Jumping occurs through a kind of handshake through time and space, with consciousness forming a bridge such that a person can end up in another parallel, alternate reality. While to the cosmos, both of "you" still exist, your awareness of who you are coalesces in one physical reality. (see: Manifesting, and Timeline Jumping)

Rabbit Holes are areas of tantalizing lines of investigative exploration and research, where researchers have little to no idea in advance what information will ultimately be found. Rabbit holes are threads of interconnected information, introducing thought-provoking and novel ideas and concepts.

Real-time Mandela Effects take place when shifts in reality occur as they are being observed or discussed. Sometimes, flip-flops can be seen real-time when looking directly at something, glancing away, and then looking back.

Reality Bubbles or reality zones are places of shared consensus reality in which particular histories of events are shared by groups of individuals—or are zones of experience unique to one individual experiencer. Also referred to as Truth

Bubbles by Roger Marsh in his book, *TruthBubble*. (see: Collective Consciousness, Download, and Smith Effect)

Reality Residue is some kind of physical evidence indicating historical events were once the way we remember, that no longer matches official current history. Such residue can come in the form of sketches, notes, music, artistic creations, or something created from a person's memories. Reality residue might show up on price stickers or computer systems, provided there was some way for a human to have entered that information into the system. (see: Residual Evidence)

Reality Shifts are observed when things appear, disappear, transform or transport and when we experience changes in time. Reality shifts range from the enigmatic (missing socks and synchronicity) to the completely astonishing and mystifying (the dead seen alive again, objects appearing out of thin air, spontaneous remissions, traveling far in a very short time). (see: Personal Mandela Effects, dimensional shifts, JOTT, jottles, and DOP)

Red Pill is the idea from the sci-fi action film, *The Matrix,* where the main character, Neo, is offered a choice between taking either a red pill, which would awaken him to see true reality, or a blue pill that would leave him comfortably uninformed.

Reincarnation is the idea of a person's spirit or soul being reborn in another form, incarnated in a new physical embodiment.

Resets refer to a series of events being experienced in a loop, where a reset point corresponds to the start and end points of a given loop. Some recognize resets as points where we can continue to revolve, or evolve. (see: Glitch in the Matrix, Looping, and Time Loops)

Residual Evidence, "residue," or "reality residue" are traces of the way some people remember things used to be, often showing up in artistic recreations and in conversations recorded in interviews. It can involve any type of Mandela Effect including: geography, physiology, quotes, books, music, movies, historical events, logos, and names. (see: Reality Residue)

Residue Hunters are those who seek evidence of previous realities in the form of "reality residue," "residual evidence," or indications that others also remember things in the same way that is different from currently recorded history and events.

Revision Technique is a technique popularized by author Neville Goddard to change the past, or to stop a buildup of negative thoughts or feelings, like a combination of emotional and energetic clearing filter, while re-imagining past, present, and future events in different ways. (see: Do-Overs)

Role Playing Game (RPG) is a game in which players assume the role of characters in a fictitious setting. Players play an active role in taking responsibility for making choices in their roles. (see: Non Player Character (NPC))

Simulation Theory suggests that we are living in some kind of construct, such as a computer simulation. This theory assumes existence of an unspecified higher form of intelligence or consciousness to be operating the simulation, and additionally assumes there must be some unspecified something that is genuine and real that is being simulated. (see: Dream)

Sliders is an experience where street lights wink out as you go past, usually while driving in a vehicle. Street light interference (SLI) phenomenon is a term coined by paranormal author Hilary Evans; thus SLIders are those whose presence is associated with such effects.

Smith Effect is also known as "Agent Smith Effect" and "Mr. Smith Effect," from the character by the same name in the *Matrix* movie. This effect is observed when people start off remembering things differently than the current official version of history, until someone indicates the current version of facts and history. Suddenly, they recant their stated memories, and say they remember how it is right now. (see: Download, Reality Bubbles, and Taking the Update)

Sovereignty is authority to govern oneself based on personal conscious agency and freedom to make personal choices. (see: Free Will)

Synchronicity, or meaningful coincidence, is a term that was coined by Swiss Psychologist Carl Gustav Jung to signify *"the simultaneous occurrence of two meaningful but not causally connected events,"* or as *"a coincidence in time of two or more casually unrelated events which have the same or similar meaning... equal in rank to a causality as a principle of explanation."* Synchronicity is the occurrence of a physical event in the world which occurs at or near the same time that it is being discussed or thought about. The essence of synchronicity is felt, for there is often significance and meaning associated with it.

Take the Update is the sometimes discernible moment when someone switches from one set of memories to another—typically the more mainstream viewpoint. (see: Collective Consciousness, Download, Reality Bubbles, and Smith Effect)

Teleportation is the instantaneous movement of some physical person or thing to another location—which might be very close by, or might be rather far away—without physically traversing the space in between.

Time Anomalies are noted by Mandela Effect experiencers observing such things as: time loops, retrocausality, time slowing down, and time slowing to a stop.

Timelines are representations of chronological arrangements of events. Timelines may be thought of as including multiple possible futures in a tree-like structure, where there may also likely be many possible pasts. The concept of timeline can also sometimes involve an assumption that we are moving through a linear progression of events, where cause precedes effect, and choices and decisions influence the way events unfold. (see: Orthogonal Time, Nonlinear Time)

Timeline Collapse refers to the idea that timelines can be combined and folded into one continuous stream of events.

Timeline Edits refer to the idea that we can influence events in history by providing positive energy such as gratitude and reverence to influence selection of sequences of events we most truly appreciate, value, and need.

Timeline Insertions of negative or positive characteristics are sometimes noted to affect historical events. Examples of this might include the Black Tom event, when in 1916, agents of the German Empire sabotaged two million tons of US-made munitions in the New York Harbor, killing four people and destroying millions of dollars worth of military goods, which was news to many people who had carefully studied WWI history, yet never heard of this event. This act of sabotage occurred in the Black Tom railroad yard of what is now a part of Liberty State Park, shattering thousands of windows from lower Manhattan to Jersey City, and even leaving pock-marks on the Statue of Liberty.

Timeline Jumping involves shifting attentional awareness from one series of possibilities—or timeline—to another. (see: Manifesting, and Quantum Jumping)

Time Loops are cycles of repeating events of a metaphysical nature, in which identical sequences of events recur, such as a person entering a building twice. (see: Glitch in the Matrix, Looping, and Resets)

Time Shifts are reality shifts in which the dimension of reality we know as time undergoes some kind of observable transformation. Time shifts appear to us in such a way that we might observe time loops, time travel, time slowing down, visits from future possible selves, retrocausality, and walks back in time.

Trigger Points are memorable moments associated with a person's meaningful memories, in such a way that inspires further action, research, exploration, investigation, or some other kind of conscious active agency.

Unity Consciousness is the idea of an underlying cosmic unity serving to unite everyone and everything. Unity consciousness can be viewed as an "I Am" presence. (see: Non-duality)

* * * * * * *

Within various branches of the Mandela Effect community, the types of Mandela Effects being recognized can be as unique as the individuals involved. The next chapter, *My Mandela Effect Life,* provides some highlights of my personal Mandela Effect experiences, with an intention of sparking insights into what we can glean from one person's subjective experiences.

~~~~~~

EXERCISE:
PUBLIC FIGURES ALIVE AGAIN

Have you had a déjà vu feeling when hearing news of a public
figure's death? Document who it was, when you first heard the
news, and other associated details. Note today's date, and whether
that person is still alive, or has been reported as having died
recently.

~~~~~~

EXERCISE:
MANDELA EFFECT TERMINOLOGY

What ideas do you see in Mandela Effect terminology that resonate
with your experiences? What terminology seems most interesting,
exotic, inviting, provocative, thought-provoking, or unusual? Are
there words and phrases you prefer to the ones mentioned here?

~~~~~~

EXERCISE:
MANDELA EFFECT CATEGORIES

Which types of Mandela Effects do you most often experience?
Make a list of categories such as: human physiology, geography,
historical events, foods, art, music, products, books, and movies—
and list some of the Mandela Effects you've encountered so far.

~~~~~~

*"There is nothing like returning
to a place that remains unchanged
to find the ways that you yourself
have altered."*

—Nelson Mandela

Chapter 3

MY MANDELA EFFECT LIFE

The Mandela Effect is ultimately a highly personal Exceptional Human Experience, capable of dramatically transforming one's worldview, so perhaps the best way to understand the Mandela Effect is to review one person's subjective experiences. This chapter highlights some of my most noteworthy personal Mandela Effects and related experiences.

A short list of some widely acknowledged Mandela Effects that have made themselves evident to me over the years include: a change in American Thanksgiving from the third Thursday of each November; the character "Dolly" having braces in the 1979 James Bond movie, *Moonraker;* the spelling of Haas avocados; the VW logo having been connected; Nelson Mandela having died twice; and Jane Goodall having been murdered within a month after Dian Fossey was killed in December 1985. I have memories of television commercials for Stouffer's StoveTop Stuffing, and Ed McMahon delivering gigantic checks to people who won the Publisher's Clearing House sweepstakes. I remember my heart beating slightly on the left side of my chest (I could feel it there), and my teachers telling me that the heart was positioned slightly to the left in our chest. I remember Laurel and Hardy saying, "That's another fine mess" (not "nice mess"); I remember that Richard Simmons wore headbands; I remember that airplane engines were positioned mostly underneath the wings (not so far ahead of the wings); and I remember the iconic line, "You've got some 'splaining to do," from the *I Love Lucy* TV show.

None of these memories currently match what has been historically documented, aside from reports from other Mandela Effect-affected individuals. These memories are especially strong for me, and I know of no mainstream reason why that should be. I'm certain I'm not mixing things up, or getting confused, so what's going on?

I have researched and written about this phenomena under the name, "reality shifts" since the 1990s, though I first noticed what we now call Mandela Effects beginning in my childhood. Perhaps due to my noticing Mandela Effects for decades before they received this name, this phenomenon seems like a normal part of the natural world to me.

INDRA'S NET

When I was about five years old and living in the Sacramento, California suburbs with my parents and sister, I had an especially vivid vision one day. I viewed all of reality stretched out as far as I could see in every direction as a shining golden web of consciousness, set with jewels of individual being. I intuitively understood that I was here in this life to assist in honoring and gently raising this web of consciousness. I later learned that this vision has been seen by others, and that it is recognized as a metaphor known as *Indra's Net,* imbuing a sense of the indivisibility of the cosmos. In Buddhist scriptures, such as the *Atharva Veda,* Indra's web or net is unbounded, and features an infinite number of faceted gemstones at each intersection or nodal point, with each gem reflected in all the others.

I remembered a sense of being pure consciousness before I was physically born in this life, at one point feeling there must have been some mistake for me to be born into this world, where people so often think one thing, while feeling something different, and then sometimes saying and doing things that are not aligned with highest self awareness. I would have given up on life altogether, if I had not experienced what might best be described as an angelic intervention in my life at age five. I felt the presence of angelic beings of light who assured me that I had chosen to be alive on Earth at at this time. I remembered existing before being born, in a blissful, ecstatic state—and this world seemed misaligned to me. I was ready to return to the between-lives state of being, yet angels reassured me that if I opted to end my life in childhood at an early age, I would immediately choose to return, and I would have lost my five year (at that time) head start. Bolstered by this news, I was delighted to choose to become the best version of myself, rather than die at a young age. At the age of five, I did not fully appreciate that my life is not just my own, and the immense value we each have to one another and the world. [3-1]

WEATHER CHANGES

Since I was quite young, I've observed reality shifts where physical reality responded to my thoughts and feelings. In a memorable moment in the 1960s, I noticed one rainy day while gazing out the living room window into our backyard that when I thought, "Stop rain," the rain would stop. When I thought, "Start, rain" the rain resumed. I did this several times, going back and forth and noticing immediate changes in the weather outside—to the point that I wanted to show this to my mother. I ran with excitement to find my mother, who greeted my announcement with a sigh of resignation. When I was unable to demonstrate for my mother what had been working perfectly just moments earlier, I recognized the power that each of us has to effect mind-matter influence within supportive "zones" or "bubbles." It seems we may exert positive influence best when we don't make a big public announcement about it. I subsequently made sure that I was quiet about any mind-matter interactions, such as any time I helped our family's car to start, or our tube television set to resume operation.

EXPERIENCING DIFFERENT MEMORIES

In the 1970s, my parents and sister and I vacationed on a houseboat on beautiful Dal Lake in Srinagar, Kashmir. My parents purchased furniture from local woodcarvers there, and made arrangements to have a set of hand-carved tables and a secretary desk shipped, in pieces, back home to California. We were in the process of building our new home at that time, and I remember my parents promising the woodcarvers that we would photograph the furniture after it was reassembled and in place in our new home, which took several months. Once the furniture was in place in our new house, I wondered when my parents would take the photographs I'd heard them promise to send. After about a month with no pictures taken, I asked my parents when they planned to take the pictures they had promised. My parents replied, "We never promised that," which shocked me, since I had heard them make that promise, and now witnessed them denying it, as if they'd said no such thing. This was a significant experience for me, since my parents are consistent, reliable, and honest and would not break a promise— leaving me with the realization that both realities were equally valid.

I HEARD A SONG BEFORE IT WAS RELEASED

As a teenager in 1977, I loved listening to the FM radio to hear new songs when they came out. I remember being confused one day to hear a song announced with much enthusiasm and gusto from the radio station's DJ, that he was about to play a great song for the very first time. You can thus imagine my surprise and disappointment when the song he played "for the first time" was the song I was quite tired of hearing. I'd heard it played in seemingly constant rotation for at least several days, if not a week or more—yet it was being lauded as a brand new song, premiering that day.

TIME ANOMALIES

I met my future self in a rather anomalous time event on the evening of June 24, 1978, when lying in my bed in a dreamy state in the dark of my bedroom. I noticed a radiantly shining woman step through the full-length closet mirror doors and into my room. I spoke to her in surprised greeting, and she responded to me telepathically. This woman looked like a grown-up version of me, several decades older than my teenage self. She floated from the closet doors over to my oak roll-top desk, where she reached down to the lower left bottom drawer and removed something, before her graceful, soothing presence returned to and disappeared back into the closet doors from whence she came. A few moments later, my father opened the door to my room, and asked if I was alright. I replied I was fine, and asked why he asked. He said he'd heard sounds of someone talking. I told him it was me talking, but omitted that I'd been talking to my future self, as I was still processing this turn of events. And I don't talk in my sleep. The next morning, I realized with a shock that letters I'd hidden under the bottom left hand drawer of my roll top desk were missing. I walked briskly down the hall and accused my sister of taking them, but she vehemently and convincingly denied this accusation, leaving me to believe the visit from my future self had been quite real.

Another time anomaly happened in the summer of 1991. I was walking across the marble floor of the Lausanne train station in Switzerland, where we lived at that time, a few paces behind my husband, who carried our one-year-old baby atop his shoulders. She had been holding onto his head and hair for support, while he carried his suitcase in one arm, and baby supplies in the other. I walked a few paces behind them, noticing that our daughter released her grip on her father's hair, and was happily waving her arms free so they rose and fell with each step he took, as she bounced along. Without warning, our daughter began to fall back, headfirst, toward the hard marble surface of the train station floor. I realized with a shock there was no way he could stop her from falling, and I was about nine feet behind—much too far to catch her as she fell. I felt a rush of adrenaline as I longed with all my heart to catch her, somehow. In the next moment, I continued walking forward at what seemed like normal—albeit fast—speed, while everyone around me moved increasingly more slowly. I observed people were moving in slow motion, slowing down so much that the sounds of footsteps clicking on marble also slowed down to deep, low, booming sounds—and then near total silence as everyone around me reached a state approximating suspended animation. I remained focused on catching my daughter, and was thrilled to easily catch up to my husband just as our daughter fell freely into my arms! As I caught her, time resumed its normal flow, and people resumed moving at normal speed.

I've noticed time loops, repeating events, and sounds of people appearing before they've arrived. This is known to Norwegians, and a Norwegian word which describes such events is *Vardogr* (vard-deh-AY-grr), meaning "premonitory sound or sight of a person before he arrives." A similar concept is known in Finland as Etiäinen, meaning a precognitive vision or dream. [3-2] Intriguingly, I am part Finnish and part Norwegian, so perhaps there may be some ancestral aspect to this!

In April 2001 while attending a Science and Consciousness conference in Albuquerque, New Mexico, I was talking with another conference participant when we both observed another woman enter the Crowne Plaza Hotel where we were staying, and walk past us—twice! The first time she walked by, we took little notice, but the second time she again entered the lobby area from the parking lot and walked past us to get to the elevators and stairs, I stopped her and asked if she'd just come through this same lobby a few minutes earlier. She said she had not, yet both my friend and I were certain she was the exact same person we'd seen walking through a few minutes earlier.

One afternoon in July 2010 while I was cooking something in the kitchen, and had the front door open to bring a breeze of fresh air into the house, I heard my daughter and her friends at the front porch, happily talking. I shouted to them that the door was open and I was in the kitchen, and for them to feel free to come inside. But then the house was strangely quiet, so I ventured out to the front porch, to see where they were, and found nobody there. A little while later, my daughter and her friends arrived, and I heard their voices once again. I asked them if they'd come and left, and then come back again, and my daughter said they'd just arrived for the very first time. [3-3]

In October 2010, I dedicated a special issue of the monthly RealityShifters ezine to the Vardogr phenomenon, "Seeing Loved Ones Before They Arrive." [3-4] In this special issue, I mention these experiences, and also some other remarkable experiences, including another time that I'd seen a woman request a business card—twice— that I include in my book, *Reality Shifts*.

ALIVE AGAIN SHIFTS

While I was a physics undergraduate student at UC Berkeley, sharing an apartment with my roommate, Kathryn had brought home a kitten named Ashes, that she raised to be a lovely full-grown cat that we adored. I became close to Ashes, who was my constant study companion while I prepared for exams during my days as a UC Berkeley physics student. A few years later, I moved across the street, and was delighted when Ashes came to visit me, spending lazy afternoons in my garden. One day I was shocked and devastated to hear the news from Kathryn that Ashes had been hit by a car, and he'd died. About a month later, I was astonished to see Ashes walking toward me in my backyard! He was dirtier than I'd ever seen him, with matted fur, and he was a bit greasy to the touch—yet obviously he was the same wonderful Ashes I'd come to know and love. I was stunned by this development, and exclaimed to my boyfriend, *"It's Ashes, and he's ALIVE!"* My boyfriend was startled by this news, and said he had just a foggy memory of something like that, almost as if he were remembering a dream that Ashes had died. I was so grateful to get to spend several more months with Ashes after that, and made sure to not take him for granted. [3-5]

In 1997, I read an article in a newspaper reporting Larry Hagman's death. He'd been a big star on the TV show, *Dallas*, playing the role of J.R. Ewing, which was a pivotal part of the buzz at the time with viewers wondering, "Who shot JR?" His death was reported on TV, in newspapers, and in magazines, and I remember seeing these reports at that time. When I later saw fresh news that Larry Hagman was starring in a new show, I found this to be extremely odd! In Autumn 1998, I discovered that Larry Hagman had not died, but rather had undergone a liver transplant—which definitely was not what I'd remembered having seen reported the previous year. [3-5]

REAL-TIME SHIFT

In 1994, my husband and I attended a workshop at the Monkey Island Hotel on an island on the river Thames, situated in the royal borough of Windsor, England, just a bit to the west of London. I noticed with joyful anticipation that there would be a speaker at this workshop named Steve, associated with the international architectural firm of Ove Arup. I found this especially intriguing, since this workshop was in honor of my husband's software program, and I was interested to hear how this large architectural firm utilized the software. I was super excited to see that Steve's talk was printed on the program for that afternoon, after lunch. At a break in the workshop, I left the room for refreshments, and when I returned, I was surprised that Steve's name was gone from the program! I looked around, and noted that his name wasn't just gone from my program, it was gone from every program in the room. I asked my husband what he remembered, and he also recalled Steve's first name—and that he worked at Ove Arup—though neither of us could remember his last name. We wondered how his name could have so abruptly gone missing from every single program in the room, and how it could be possible that all the program sheets could have been re-typed so quickly at the break and put back exactly in the unique places where each participant had left them. Clearly, no programs were retyped, and we had experienced a reality shift.

BILOCATIONS

People have observed me bi-locating on a couple of occasions, and in both cases, I was daydreaming at the time. One cold morning in the 1990s, I stayed in bed, daydreaming that I was walking down the hall, flipping on the light switch in my daughters' bedroom, opening a window shade that was far too heavy for them to operate, and asking my daughters to wake up. A few minutes after daydreaming all of this, I was surprised to hear my daughters talking with one another, since they just about never woke up for school by themselves. I then heard them coming down the hall to my bedroom, where they asked me, *"What are you doing back in bed?"* I asked what they meant, and discovered that during that exact same time I'd been daydreaming under the covers in my bed, my daughters had seen me come into their bedroom, switch on the light, open the window shade that was too heavy for them to lift by themselves, and tell them it was time to wake up. If the fact that they were awake without me having awoken them wasn't enough, I could also see that their light switch had been flipped on, and their very heavy window shade was open. I've heard skeptics muse that possibly I might have been sleepwalking the morning that my daughters wondered what I was doing back in bed—but that would have been unusual, since I don't sleepwalk. That would have been the first and only occurrence. And sleepwalking can also be ruled out in another bi-location experience, where I was seen miles away from where I was at the time.

In 2009, I spent one sunny weekend day indoors at an author event. It was such a gloriously beautiful day, that I daydreamed while gazing out the windows that I was actually outside, walking along Solano Avenue in Albany—several miles away from where I was at the time. When I got back home, later that day, I saw that my friend Katrina emailed me, to let me know that she'd been surprised to see me walking down Solano Avenue in Albany, California since she was fairly certain that she'd heard that I'd be at an author event in Pleasant Hill, California that afternoon. I asked her when she'd seen me walking down Solano Avenue, and learned she had seen me at the exact same time that I had been looking out a window at the book event, daydreaming that I was walking down Solano Avenue!

TIME AND SPACE

A remarkable shared "Aha!" Mandela Effect moment occurred one day in the 1990s, when I met two friends for brunch, and we later walked along the Berkeley Marina, as was our custom. I mentioned some reality shifts I'd recently seen, and said that I wished that my friends could see a reality shift, too. At that very moment, one occurred, right on the spot! Cliff asked Jan and I, *"Do either of you recall ever seeing that statue here before?"* as he pointed at an enormous concrete sundial sculpture. Jan replied, *"No ... I don't."* I was thrilled, because I had sometimes seen that sundial sculpture there before, but never with my friends! We were certain that this sculpture had not been there when the three of us had walked after one of our brunches before, because it now obscured the view of a fascinating Guardian sculpture: a larger-than-life size Asian archer on horseback with fully drawn bow, aiming his arrow out over the bay—shooting over the sundial. There was a plaque on the sundial, stating it had been a gift from the city of Sakai, Japan to Berkeley, California in 1970, and it indeed appeared to have been there a long time. Yet we were certain it had not been there when we had been at the Berkeley marina together before, when we walked through a large, open plaza just below the archer statue, only a few months before.

A fascinating footnote to my sundial reality shift encounter is that it closely parallels an experience described by Christopher Anatra,

President of NECS, after he saw that a house on his daily commute suddenly featured a prominent sundial that had never been displayed there before. Anatra stopped to inquire if that sundial was new, and learned it had been there for some time—even though he drove that route regularly, and had never seen it there previously. The symbolic message of a sundial to be mindful of space and time feels like a meaningful clue that the cosmos brings to our attention. [3-6] Anatra pointed out in his interview with Matthew Harris the same thing I mention in my book, *Reality Shifts,* and also my article about the top ten ways to shift reality: the more observant we are, the more likely we are to witness Mandela Effects and reality shifts. [3-7]

MUSCLE MEMORY SHIFT

I was convinced of a personal reality shift when I utilized my usual muscle memory to turn a knob on the timer dial on my clothes dryer. I had been operating this dial regularly for years, and I could have started it in complete darkness, since I made the same identical hand movement so often. One October evening when I began to set the timer for twenty minutes, I was stunned to discover that the dial no longer turned the same direction or from the same position. Instead of turning the dial from four o'clock to five o'clock, I now needed to turn it from about eight o'clock to about seven o'clock. This movement required both a completely different position and rotation than my muscles were accustomed to, so I stopped and stared at the timer dial for a long time, to figure out what to do to obtain desired results. [3-5]

HEALING PEOPLE, ANIMALS, AND THINGS

I've witnessed numerous experiences of instantaneous healing for myself, friends, family, loved ones, a beloved pet cat, and our family dog. The Mandela Effect and reality shifts are closely related to what some would call miraculous healing, and much of my book, *Reality Shifts*, is dedicated to covering such experiences.

Reality shifts indicate that miraculous change is possible, and might be happening around us all the time. Reports of spontaneous remissions of cancer, surprising awakenings from comas, people who can lift a heavy car off of someone in an emergency, and other amazing things are sometimes reported in the news. What we tend to overlook is the possibility that the ability to transform this world lies within each and every one of us. We are each creators in this universe, and the very act of observation can change reality. We are capable of healing not just other people, plants, and animals with our feelings of love and thoughts of well-being, but also "non-living" things around us.

I had an interesting discussion with my literary agent about my use of the phrase, "healing things." She disagreed with me that things could be healed, since they are after all, things. *"Yes, but I feel consciousness exists in everything, and that everything responds well to love,"* I explained. My agent cut in with her brisk New York accent to repeat firmly, *"Things aren't healed."* I confirmed that I knew what she was talking about by replying, *"You mean things can be fixed, but not healed, according to the way most people see the world."* "*Yes!*" she replied, brightly.

I was amazed by this conversation, because I'd been trained to imagine seeing the world as an impartial observer throughout my college education, yet I'd secretly wondered how the numerous revolutionary implications of quantum physics could have gone completely unnoticed by the majority of my professors. My quantum physics classes thrilled me in a way classical physics had never done, by showing me first-hand how it was possible to "go by feel." Rather than certainties, quantum physics dealt with probabilities. Instead of being uninvolved observers in quantum physics lab experiments, quantum physics described how the very act of selecting observational perspective influenced what was observed. Rather than matter being all that matters, there is something deeper and more mysterious going on. Quantum particles separated by great distances constantly influence their non-locally entangled twins. They do this so dependably, and to such a degree, that physicists and engineers are developing quantum computers to harness this "spooky action at a distance"—which was Albert Einstein's pet phrase for quantum non-locality.

My experiences observing things healing have been similar to observing people healing. In both cases prior to healing, I sense energy imbalances. In both cases, by feeling love for the person or thing to be healed and bringing healing energy in, I have often witnessed healing reality shifts. It's easy to forget to try energy healing first, when we've been raised in 20th century modern culture with it's assumption that physical material intervention is the best course of action. I often start out doing things the traditional way, by taking broken equipment in to the repair shop—sometimes later remembering that energy healing often works best to fix things that repair shops can't.

When my Epson color printer started printing splotchy bands of color on every page, I took it to a competent printer repair shop. When the first repair didn't fix the problem, I brought it back to the shop, and when I came to retrieve it the second time, I was told by the technicians that my printer was beyond repair. They said the connections had been checked and cleaned, ink had been replaced, jets had been cleaned—and every mechanical part of the system had been fully scrutinized and reviewed to the best of their ability—and they declared the printer was defunct. Their recommended best course of action was to purchase a replacement printer, and discard the broken one. I returned home, feeling sad. I felt how much I loved my printer and how much I wished it was operational, but didn't yet remember energy healing. After a couple of days, I went shopping for a replacement printer, and returned home with a new one. No sooner had I set the new color printer box down in the same room with my old printer, than I discovered my old printer was fully functional! My printer worked perfectly well for years after that. It seemed to me that my feelings of love for my printer, combined with a neutral, detached state of mind provided an optimal healing environment for my printer to return to full functionality.

My experience healing my printer came to mind when I visited a friend who has a scanner, and asked her to please scan a Polaroid aura picture of me. I showed her the photo with its beautiful rainbow colors superimposed over me. My friend admired the photo, looked a bit flustered and embarrassed, and replied, *"My scanner isn't working, and hasn't been working for months now. Everything it scans comes out looking green. I've checked all the connections, and done everything I could think of, but it's still not working."* She then showed me a scanned family photograph with smiling green people and a green background. I toyed with the idea of joking with her about her Martian relatives, but thought the better of it. Instead I said, *"I think your scanner can work, and I'll gladly help you heal it, if you want. There's nothing to lose, and if we heal it, you'll have a working scanner again!"* I was happy to see her eyes brighten, and glad that she was up for giving it a try. I asked if she could feel or see the energy around her scanner—and told her that I saw a dark energy that felt heavy and thick as I moved my hand through it. I indicated for her to do the same, and she observed that the area around the scanner felt noticeably different from the surrounding areas. We then felt how much we loved the scanner, called with gratitude upon the healing spirit of the scanner (the "scanner angel") to assist us in healing it—as I gave my friend a number of options for ways we could bring in healing energy. She selected my personal favorite, the "love blast," and we proceeded to clear out the sick energy and replace it with positive, healthy energy. I could feel the blast of shining love radiating from the scanner blowing all the dark stuff away, and did a healing/cleansing on all the connectors and cables before turning it on, and seeing it work—perfectly! [3-8]

COIN BENDING AND SYNCHRONICITY

After meeting Uri Geller and having seen him bend one of my spoons in 1999, I felt inspired to hold a spoon-bending party at my house. At this small gathering of friends, I was amazed to see my daughter looking relaxed and energized, gazing out the window into the garden with a fork resting on her wrist. While her eyes focused outdoors, the fork on her wrist slowly melted down, like chocolate melting in warm sunshine. She didn't move or touch the fork—while it slowly "melted" in a downward curve, when she gazed out the window to the garden.

This memory came to my mind one day when I'd parked my car at the parking lot for our local Bay Area Rapid Transit (BART) underground subway, on my way to see an event hosted by the Institute of Noetic Sciences (IONS) in San Francisco. I'd brought exact change to pay my fare at the kiosk, with just enough time to purchase a ticket and catch my train. As I inserted the last coin through the coin slot, I noticed with chagrin that this quarter was becoming soft and malleable, like a chocolate melting in warm sunshine. With such softness and a fresh curve to its shape, I could not fit it through the coin slot. I heard the "whoosh!" of my desired train arriving underground, and ran back to my car, to get extra coins this time! Once I'd paid my fare and boarded the next train for San Francisco, I saw a man seated on the train, reading a book written by Edgar Mitchell, the founder of the Institute of Noetic Sciences (IONS). Realizing I was experiencing a wonderfully meaningful coincidence, I sat near him, and we chatted about a plant consciousness experiment he was proposing to IONS, about human-plant communication. Thanks to that conversation, I was able to see his experiment in action when I visited IONS later, and wrote about it in my book, *Aura Advantage*. [3-9]

When we arrived at our downtown San Francisco station, I gave my bent quarter to the first musician I saw playing music, since I was feeling so annoyed by that bent quarter. I later realized that this experience that started with my melting quarter and ended with me being on just the right train to learn about something I'd not have heard about otherwise is a wonderful example of how even seemingly frustrating events can be working out in our favor, when viewed from a higher level of self. I now appreciate the incredibly beautiful orchestration of events that had occurred that day, and am grateful to have gained a glimpse into higher levels of conscious agency at work.

VISION OF TREES APPEARS IN SKY

When we maintain a state of unconditional love and choose to feel joy and reverence, it's sometimes possible to see things that are invisible to ordinary everyday vision. I witnessed one of the most striking cases of this in 2004, when my young daughter asked to visit two trees she loved that we could see miles away from our home, up high on a distant hill. We rode our bicycles out near to where the trees were, and when we got closer, I was surprised to see that a forest blocked our view. I could see the forest from the bike trail, but not the two distinctive trees we had ridden uphill for an hour to visit. I'd been feeling confident that we'd be able to find these special trees, despite not having a map or directions, because we'd been loving those trees from a distance for so many years, that I felt certain we'd be able to go right to where they were. I then saw a vision of the two trees that my daughter wished to see, up in the sky, at an angle and with the perspective that I might have expected to see them from where I then was, if the forest of trees wasn't there. This was a very different angle than what we usually saw from a distance, but this new vision looked exactly like the right trees, and seemed to indicate their location. Excitedly, I told my daughter, *"I know which way to go!"* The two trees were indeed situated exactly where I had just "seen" them to be! And sure enough, this vision provided me with the ability to make a direct path off the trail and through the forest, where we arrived at the trees we'd come to visit.

FLAT TIRE FIXED BY GOD

I was mulling over what matters most to me one day in July 2015, appreciating how I feel certain that every good thing comes from God, and feeling deeply appreciative of how health, wealth, happiness, luck, and wisdom seem empty to me without Spirit—without God. I was driving my car while feeling a meditative sense of, "How good can it get—how God can it get?" reverence, when I heard a sudden *"Pow!"* noise, followed by intermittent hissing and a *"Thwup-thwup-thwup"* sound as the steering pulled to one side, and the front right side of the car went slightly down—all the distinctive sensations of my car's right front tire going flat.

"Oh, no," I thought, as I found a clear place to pull over and park. I got

out of the driver's side door and walked around to inspect the right front tire. What I saw confounded me—the right front tire looked good. I walked back to the driver's side door, and then thought to myself, "*No way!*" returning to take a second, much closer look this time. The tire appeared to be completely inflated, with no indication of any type of trouble. From what I'd just felt and heard, I expected to see a flat, deflated tire on the right front wheel. But the tire looked good. I returned again to the driver's seat, sat down behind the wheel, started the engine, and slowly pulled back onto the road, listening with windows open for any unusual sounds. With heightened sensitivity, I heard the gravelly noise of tires crunching over asphalt—the normal noise of driving down the street.

I checked the tire at the parking garage, and again when I got home. Every time I looked, I saw that it was fine. As the flat tire had been somehow repaired before I'd seen it, the memory of what I heard and felt seemed more distant. I began to doubt that I'd really heard and felt a flat tire, but I also had a nagging suspicion that that tire might actually somehow still be flat, and its air pressure should be checked. I asked my husband to please check the tire pressure on the car, when I returned home. After all tires were checked, I was stunned to hear that one tire was fully inflated to manufacturer's specifications. I asked my husband, "Which tire is fully inflated?" and he replied, "The right front tire." Wow! The other three tires were all slightly deflated, and now it seemed I had some kind of evidence that something had definitely happened with the right front tire.

One might think I'd be properly wowed by all this, but instead I found this reality shift almost too miraculous to wrap my mind around. I thought to myself, "*You must be confused. You must be mistaken. No way could something like this happen—it's just too improbable. I've never heard of such a thing. Therefore, it's much more likely that for some bizarre reason I mistakenly heard and felt indications that seemed to suggest my car was experiencing a flat tire, when it actually wasn't.*"

The next day, I learned that my phone answering machine was full of messages—dating back to 2012! I had been archiving some messages from friends and family before erasing them from the answering machine, and two months earlier I'd just transferred the phone recordings before deleting the messages. So how could the machine be full of messages now? I found my cassette tape and recorder, and froze in confusion and astonishment as I held them in my hands. The batteries in my tape recorder were old and starting to leak, the cassette tape cases were dusty, and my handwriting on the cassette indicated I'd last archived phone messages in 2012. After the batteries were replaced in the tape recorder, I heard a recording of my own voice ending the last recorded phone messages stating the year was 2012. *"What?! This doesn't make sense—I just transferred phone messages a couple of months ago!"* I had replaced the batteries, dusted all the cassette cases, written notes on the cassette, recorded my voice with the year 2015—but now there was no record of any of that.

After finding a new cassette tape to record on, with the message transfer once more underway, I wondered, *"What on Earth is the point of having to do all this work over again?!"* I then remembered how I'd disregarded the miracle I'd just experienced with my instantly-repaired flat tire. While it is true that car miracles (such as traveling farther in less time than should be possible) tend to be more common—probably partly due to people being in relaxed, meditative states of mind while driving—this car miracle was different. At the time my right front flat tire went flat, I'd just started driving and was engaged in heartfelt morning prayers. I was feeling about the highest level of reverence and rapture I can sustain while driving a car on that particular morning. This is probably the most noteworthy detail in this miraculously-fixed-tire experience, since it felt to me like God fixed my flat tire. Instantly. With perfect inflation to exact manufacturer specifications.

It had been difficult for me to accept the miraculously-fixed-flat-tire, so I could see the advantage of having gone through the frustration of having to transfer phone messages all over again, while observing how my methodical record-keeping was so different than expected (with notes on the cassette tape showing 2012 rather than 2015, voice-over recording of the 2012 date, and placement of the tape after the last transfer of messages, with dust over everything as if it had been sitting undisturbed for years). All of this really drove home the point that there is no limit to how big or fast things can change. This includes instantaneously fixed flat tires! Especially when we're thanking God for everything good.

DO THAT THING YOU DO

After enjoying a holiday meal at my parents' home over the holidays one evening, my sister and I had just finished loading the dishwasher with dirty dishes. We pushed the "start" button on the dishwasher, and instead of hearing a happy sound of an operational dishwasher, we heard a "clunk" sound and I smelled something like oil burning. My sister and I looked at one another with concern, realizing that this would be a bad day for the dishwasher to break down. We had memories of a not-so-distant holiday when our parents' dishwasher had died, and we'd ended up doing dishes by hand for hours. *"Do that thing you do,"* my sister implored, and I smiled with surprise. I felt a sense of giddy joy that she'd asked, which was wonderful, since it immediately lifted my spirits into just the right kind of energetic sweet spot for success in this endeavor. I next described an overview of the required steps, starting by saying it's a good thing we're now smiling and laughing, since this kind of strong, joyful emotional energy is optimal for best results. I also explained that primarily what we'll do is hold the thought in mind that the only acceptable possible realities are those in which the dishwasher is running well, while ignoring the "clunk" and grinding sound we heard, and the smell of burning oil. We'll then relax and let go of all concern, and when we feel relaxed and ready, we'll push the "start" button, while envisioning it running just fine. Which is precisely what happened next.

INSIDE THE MANDELA EFFECT

It's one thing to know that superpositions of states can exist involving many possibilities of various events having occurred—but it's something else entirely to be the one who is existing in a state between life and death, like the cat in the Schrodinger's cat experiment. The imaginary cat in the Schrodinger's cat experiment is locked inside a box with a vial of poison that could be triggered randomly at any moment. Because the vial of poison is broken based on the quantum radioactive decay of a quantum triggering device, this imaginary cat can be considered as being hypothetically both alive and dead at the same time. I now have a better idea of how that cat might feel, after having gone through a similar experience.

I'm now in the remarkable position to be able to report from inside the Mandela effect. I've been reported dead, and alive again. In early 2017, I survived a cold that moved into my respiratory tract and went into pneumonia, bringing me close to death. I only shared this with immediate family, so as to not unnecessarily upset people. After my recovery, I gave the subject no further thought.

In April 2019, I received an email from Steve Boucher, informing me that strange as it may seem, he remembered seeing reports of my death. When I wrote back to inquire what details he remembered, he shared that: (1) the time of my death was early 2017, (2) the cause of my death was related to my having been very sick for many months, and (3) news of my death was reported through my Facebook page. All of these details matched what likely would have transpired, had I died. [3-10]

I received emails in May 2019 from people expressing gratitude that I was still alive, after sending that month's issue of *RealityShifters*, "Feeling Grateful to be Alive." Some people noticed I'd been absent. Jo wrote, *"I had thought things were very quiet for a while, as if you weren't 'there'. It was noticeable, but beyond that I didn't get thoughts that you'd passed."*

I heard from Megan who *"absolutely remembers the news of your passing (except it was posted on your YouTube channel)"* at a different time than Steve Boucher recalled—not in early 2017, but in early 2019. Megan elaborates:

> *"When I saw the announcement on your YouTube channel that you had passed it was a long note from your husband talking about your long illness and there was also a picture of you with several of your friends standing in a group. There also was a memorial video that was from your YouTube channel with music and shots of you throughout the years! The weird thing is that it was only maybe 4 or 5 months ago, when I was going through my own personal crisis of sorts. It was a strange time for me as I felt very close to death myself. You mentioned that if I notice your timeline shift then maybe I had one myself? I have an incredible circle of friends and support and I wonder if they helped me choose to stay in this timeline even though I was at that time very ready to go? I have told your story of the timeline where you passed to several people since it happened. It really affected me, and created a feeling that has been with me since—a magic of some sort, electricity kinda—I can't explain it, but I bet you understand exactly what I mean! The other thing worth mentioning is that your YouTube channel looks very different than the one from the other timeline, and also you look slightly different too! In the other timeline you did not have a close up picture of you on each of your videos and the lettering style was different as well. I almost didn't recognize your videos here (and actually didn't at first)."*

Stinne in Denmark wrote that she found my website after watching a YouTube video of Rupert Sheldrake talking about me and mentioning that I'd died. Stine felt sad that she'd missed out on contacting me.

When she followed up on Sheldrake's mention of my website, she looked up my work and was delighted to discover that I was alive! When I inquired which Sheldrake video she'd seen that had mentioned me, Stine said that the video was gone without a trace.

Laurie wrote, *"I happen to be one of those who was on the timeline when you died. I remember being very sad as the female voice in the quantum physics realm was no longer on the planet I was part of. I was on your email list for your newsletter. I received an email from your husband letting us know you had passed. I had Deepak to guide me through newer understanding, but I truly wanted a female scientist. (I also experienced the "loss" of Jane Goodall and Dian Fossey.) Glad I am not on that timeline any longer! So very happy your light is still shining and so many can see it now."*

This has certainly been a mind-expanding, incredible experience for me, as well as a continuing opportunity to glean insights with regard to how it feels to be "inside the Mandela effect" as an Alive Again survivor. I now recognize that someone who feels they're about to die, but then miraculously survives (sometimes against all odds—such as "tunneling" or "teleporting" to the other side of an oncoming truck) is likely to have reached a major "level-up" opportunity in life to become more fearless and more unconditionally loving. I reached such a point in my life starting at the time I almost died in early 2017, and I am certain that decision point continues to make a huge difference in my life. [3-11] I have a clear sense that I made a jump to live—assisted by all those observing me healthy and alive now. I had a sense when surviving my close brush with death in the winter of 2016/2017 that I was choosing a new-and-improved version of me, as if a more fearful possible version of me had died at that time. This brings to mind recent news of how physicists can predict the jumps of Schrodinger's cat—and finally save it, since we now see scientific evidence indicating that future observations can influence past events. [3-12]

DELIVERY PACKAGE SUPERPOSITION

I sometimes witness evidence of superposition of states in the physical world, such as the time in December 2018 that I once received multiple simultaneous delivery status email messages from FedEx, informing me of where the item I had ordered was currently located, and when it was expected to arrive. I had been wondering what had happened to the handmade walking stick I ordered, since it had not yet arrived. I opened each of the delivery status emails, and read reports that according to three out of four of them, my walking stick had already arrived and been delivered to me.

Since I had not yet actually received my package, I refrained from becoming anxious or upset. I knew from experience that whatever possible outcome I focus most energy on tends to be selected, so I focused attention on the single email message indicating that my package was en route, and expected to arrive within the next couple of days. Despite the fact that 75% of the email messages I received appeared to indicate my package had already been delivered, and might subsequently have been stolen by a package thief, I remained relaxed and optimistic. And I was delighted to find within a day or two that my walking stick package indeed arrived just fine. I'm certain it helps that instead of having some part of my rational mind believing that such superpositions of states are impossible, I instead remember scientific news reports such as how a recent physics experiment challenges objective reality, so I allow that observers can be at the same place at the same time, and witness reality differently. [3-13]

TWO OF MY BOOKS CHANGED

In 2019, while preparing my keynote talk for the upcoming Mandela Effect conference in Ketchum, Idaho, I wanted to reference what I'd written in my book, *Quantum Jumps,* about the meaning of the cover illustration, where I'd included an explanation of how we might make a jump to a parallel possible reality via Einstein-Rosen "wormhole" bridges, that exist as miniature black holes. I searched unsuccessfully through a paperback copy of my book, and then through the digital file, wondering what had become of that important explanation. I checked three places in the text where it logically might be, and then searched the entire document cover to cover. After a few hours of searching had elapsed, I realized that the explanatory paragraph was missing from my book! I mentioned this in my talk at the conference, and I felt relieved that people could understand that this kind of thing can happen.

In 2022, I discovered that another book I'd written had changed. I attempted to find a section in my book, *Aura Advantage,* that used to inspire vegetarian readers to send several emails to me each year. The passage in question had to do with how my salad "spoke" to me, when I'd said a blessing of gratitude and appreciation for it, before eating that day. Many people were upset to hear about my conversation with sentient leafy greens, and each year I'd correspond with people I'd inadvertently upset. The reason I was searching for this material in 2022 was because two people emailed me with topics that seemed remarkably related, and I thought it would be a simple matter to share that section of my book with them. And it would have been simple, if that part of my book still existed! I only spent about an hour on this quest, since I remembered I'd been through this with *Quantum Jumps* before. I searched the two paperback editions of *Aura Advantage,* as well as the electronic version, and couldn't find this passage that had sparked so much correspondence with readers in past years. It has been about seven years since I stopped receiving emails about this. Now that the previously offensive section is gone, without my having done anything to revise the book, I'm glad my book is no longer upsetting people.

MASK SIGN VANISHES

When I arrived at the doctor's office on January 9, 2020, after checking in with the receptionist and sitting down, I noticed that everyone in the waiting room was wearing medical face masks. Several people were coughing, and I noticed there was a sign with huge lettering by the check-in line that read, *"Everyone MUST wear a mask."* When I first arrived, there had been no line, so I'd gone directly to the receptionist's desk without seeing the sign. This sign seemed unnecessarily bossy, but with so many people coughing, I decided that wearing a medical mask was probably prudent, so I donned one of the blue facial masks. At that time in January 2020 in California, there had not yet been widespread news of the coming coronavirus pandemic, nor had there been widespread use of facial masks at that time. A few minutes later, another couple of women came in, signed in with the receptionist, and sat down. I noticed that neither of them donned a mask, and thought maybe I ought to point out the sign with giant lettering. I looked over to where it had been, and saw that it was gone! When I went in to see my doctor a while later, he asked me, *"Why are you wearing a mask? You're not sick."* I didn't know how to explain that there had been a sign in the waiting room that adamantly stated everyone was to wear a mask —because it vanished and had not returned by the time I was called to see the doctor. A while after my visit to the doctor, I felt grateful that I had seen that pushy sign, since some really bad colds were going around. [3-14]

A ROSE APPEARS, TWICE

A truly magical reality shift occurred in January 2020, while I was having dinner with a dear friend at the beautiful Dushanbe Tea House restaurant in Boulder, Colorado. This building was moved piece by hand-carved piece from Tajikistan, and is a work of art with magical atmosphere. At dinner, my friend and I had been talking about the way the Mandela Effect is providing evidence that global miraculous changes can happen instantaneously. During this blissful conversation, we talked about co-creating heaven on Earth, and continued talking long after all other diners had left. The wait staff were very hospitable, insisting that we, "*Stay as long as you like!*" So we stayed a while longer, until agreeing it was time to leave. As I moved my chair to give Nicole a hug, a beautiful tiny pink rose fell down out of nowhere. It was fresh-cut and gorgeous—and I was stunned. We had been talking about the positive vibe at the Idaho Mandela Effect conference, and the topic of choosing positivity over negativity, when this gorgeous little rose arrived out of the blue. We looked around in all directions, seeking other little roses, but there were no flowers to be seen. We were the only diners in the restaurant, and the wait staff were out of sight in the kitchen area. I've seen several things manifest out of thin air before, such as a half gallon of milk inside the refrigerator, car keys that had been locked in the trunk of the car, a tiny tooth my daughter had lost at school that day, a small crystal when I'd just told a woman she had a crystalline quality to her heart, and dollar bill after dollar bill arriving inside my wallet. [3-15] But this was my first fresh-cut flower! [3-14]

When I returned home to California from Colorado, I was unable to find the sweet tea rose that had so majestically landed at my feet. A couple of months later, in March 2020, I was astonished to find the missing rose dried in perfect condition, unbroken and resting at the bottom of the empty purse I used most every day. There's no way that fragile rose could have survived in such excellent condition if it had been there for the past couple of months—and since this was my daily purse, I was putting things into it and taking things out of it regularly. The arrival of this rose felt like a beautiful sign of beauty, peace, love, and hope.

LAUNDRY DOES ITSELF

I experienced an amazing reality shift on Groundhog Day 2020, when I was a guest on the Ripon Rabbit show about the Mandela Effect, hosted by AJ, "the Ripon Rabbit," and featuring YouTuber moneybags73, aka Evan Matraia. [3-16] Before the show started, I ran two loads of laundry; one had just finished washing and was in the washing machine, and the other was waiting in the dryer. I'd planned to complete the radio show and then move clothing to the dryer and run the dryer. But as soon as the show was over and I said "goodbye," I heard the dryer running in the laundry room. I opened the dryer and found that both loads were in the dryer, and they were dry! All of this was technically impossible, since I'd left the washer and dryer turned off during the taping of the show, to minimize background noise. I had just discussed how I love experiencing new kinds of reality shifts as long as they're enjoyable—and it certainly was wonderful to discover that my laundry had dried itself!

LONGER SILK CURTAIN

In January 2021, I noticed that a silk window curtain which I'd washed in hot water over a decade earlier was suddenly drooping down so low that I needed to raise it's curtain rod. That silk curtain had shrunk about 1.5 inches after I'd made the mistake of washing it in hot water, so I had to lower the curtain rod, leaving about an inch and a half gap at the top. Before the curtain suddenly returned to its original size, I'd been discussing with my husband how we really needed to replace that curtain. And then, a couple of days later, I noted with some annoyance that the curtain drooped down way too far at the bottom. I suddenly realized, "Wait a minute! That's the curtain that's been too short all these years—and now it's exactly the right size again?" Yes, indeed! I moved the curtain rod up, and our curtain is miraculously restored to its original length, without needing to be replaced.

TELEPORTING BOTTLE

In March 2021, I experienced a real doozy of a teleportation. I'd just finished listening to an Institute of Noetic Sciences (IONS) event, featuring Dean Radin, where he'd been discussing results from his Intentional Chocolate experiment. I was contemplating mind-matter interaction, when I walked into the kitchen, closed the door behind me, and shook the bottle of Howard Feed-N-Wax furniture polish I was carrying in one hand. Much to my surprise, the bottle completely vanished! I looked for it where I expected it would have logically gone —toward the oven—following the trajectory of my arm, and even looked inside the oven and the trash can, but it was gone. After I'd had time to appreciate that this vanishing object event was rather spectacular, I felt a sense of resonance with that bottle of furniture polish, and sent out an emotional/thought message that if the bottle wished to return, I'd be delighted, and there'd be no questions asked. A little while later, I walked out of the kitchen, across the hall, and into the living room, where I saw the bottle of furniture polish on the floor in front of the TV set!

DOCUMENTING TIMELINES

It's helpful to keep a journal of reality shifts and Mandela Effects, so you can review your experiences, and recognize patterns that become obvious with the passage of time. Those of us who notice alternate histories can help sort things out when events start seeming less than linear by keeping track of noticing three dates with their associated events/observations: X, Y, Z to accompany three important Mandela Effect moments on a timeline. For example, I would write down today's date Z when documenting a particular Mandela Effect observation that I first became aware of on date Y, involving something that happened earlier in time, back at X. If I record my own unique memory of American actor Larry Hagman being alive again, that I wrote about in my book, *Reality Shifts,* might thus be noted as:

Z: 1999   I documented my Larry Hagman alive again experience
Y: 1998   I saw news that Larry Hagman had a liver transplant
X: 1997   I saw a news article reporting Larry Hagman had died

These three dates are sometimes spread farther apart, and sometimes closer together, and they are completely unique to each of us. The date that we record the event is also an integral part of the entire experience. This kind of XYZ timeline record-keeping can be especially useful when tracking Mandela Effects that go back and forth, flip-flopping between two or more states. [3-17]

* * * * * * *

The next chapter, *History of the Mandela Effect,* delves into a chronicle of how the Mandela Effect became better known when people began sharing reports of their observations around the world.

~~~~~~

EXERCISE:
RECORD ALTERNATE HISTORIES

You can document Mandela Effects by consistently recording your experiences. Think about a recent Mandela Effect, and write down the date "Y" when you first noticed something seemed different, and also the earliest date involved, "X" for this experience. Also note the date, "Z" when you are recording this experience.

In addition to noting what you observed changing and when you noticed changes, you can add any thoughts or feelings you were aware of at those times, with any memories, sensations, feelings and emotions, as well as surrounding events and insights.

~~~~~~

EXERCISE:
PRACTICE CALM SUPERPOSITIONS

A fun way to experience flip-flopping realities is to have already envisioned facing such situations and picturing yourself staying focused on the reality you most desire. For example, you can imagine receiving email notifications that something wonderful is on the way. Picture yourself staying calm, energized, and positively focused on receiving what you most love and need.

~~~~~~

EXERCISE:
PLAY WITH REAL-TIME SHIFTS

Next time you don't get a full night's sleep, tell yourself that you got an excellent night's sleep. And next time you feel like you might be starting to come down with a cold, tell yourself that the symptoms of being sick are on their way out, and you are nearly fully back to excellent health. In either of these real-time shifts, the key to success is choosing to believe that you are in the new reality where you've had enough sleep, or are just getting over a cold.

~~~~~~

~~~~~~

EXERCISE:
DAYDREAM INTO AN ALTERNATE REALITY

Make some time to daydream about how good your life can get, and imagine you can walk between realities by walking down the hall past various doorways, and then enter a different room (or re-enter a room), imagining that you're walking into a new reality.

~~~~~~

*"What counts is not
the mere fact that we have lived.
It is what difference we have made
in the lives of others
that will determine the significance
of the life we lead."*

—Nelson Mandela

Chapter 4

HISTORY OF THE MANDELA EFFECT

The Mandela Effect is named after Nelson Mandela, who many people remember having died in the 1980s, and then surprisingly be alive again a couple of decades later. This particular "alive again" type of reality shift has been noted many times in the past. One of the world's most famous "alive again" incidence involves American author Samuel Langhorne Clemens, also known as Mark Twain.

Twain was 61 years old and living in London in June 1897, when a *New York Journal* news correspondent, Frank Marshall White, came knocking on his door. Frank showed Mark Twain two telegrams from his editor. One telegram's terse message read, *"If Mark Twain dying in poverty, in London, send 500 words."* The second read, *"If Mark Twain has died in poverty send 1,000 words."* Biographer Ron Powers points out that at this point in time, Mark Twain was creatively and emotionally exhausted, and living in self-exile, while laboring to honor the speculative debts that had financially bankrupted him. He was grieving the recent loss of his beloved eldest daughter, Susy, who had died of meningitis months earlier, and was sad that his wife, Olivia, was despondent and frail. Twain was growing tired of lecturing and writing on demand, with contractual terms so demanding and restrictive that he had recently declined to sign a contract with Century magazine that specified his obligation to write twelve humorous articles. [4-1]

The *New York Journal* published this response by Mark Twain on June 2, 1897 that read,

> *"I can understand perfectly how the report of my illness got about, I have even heard on good authority that I was dead. James Ross Clemens, a cousin of mine, was seriously ill two or three weeks ago in London, but is well now. The report of my illness grew out of his illness. The report of my death was an exaggeration."* [4-2]

News of this remark catapulted Mark Twain out of relative obscurity and back onto the world stage, with the public delighting in this refreshing—if dark—dose of Twain humor. Variations of Twain's responses have become immortalized in such shortened phrases as, "Rumors of my death have been greatly exaggerated."

This case is especially intriguing because it provides us with an early example of a celebrity "alive again" experience from over a century ago. There may have been thousands of other celebrities thought to have died before actually doing so for generations, but not until the advent of the internet could people readily share such memories.

CARL JUNG'S CURIOUS MOSAICS

In 1932, Carl Gustav Jung had an extraordinary experience while visiting Ravenna's Baptistery of the Orthodox with Toni Wolff, which he later described as *"among the most curious events in my life."* [4-3] Carl and Toni viewed four distinctive early Christian mosaics, which Jung discussed at some length during their visit to Ravenna, as Toni Wolff listened while they viewed the mosaics. Shortly thereafter, Jung and Wolff were informed that the mosaics they'd thought they had seen did not exist. Jung spoke about this experience several times over the years, such as in this 1957 account:

> *"C.G. Jung had spoken to Toni Wolff about the Baptistery of the Orthodox in Ravenna where Galla Placidia was buried. Early in the fifth century, after surviving a stormy sea voyage to Ravenna, she had built a church there in fulfillment of a vow. The original church was later destroyed, but her tomb is there. After visiting the tomb, they entered the Baptistery. It was filled with a bluish light, though there was no artificial lighting. C.G. looked round the building and remarked to Toni, 'Isn't it curious? Here are these beautiful mosaics on the west, the east, the south and the north in this octagonal building, and I can't remember seeing them before—it's most remarkable for they are so striking!' In the centre was the font; it was big for it was used for immersion. For twenty minutes they studied the mosaics. C.G. described them as about twice the size of a tapestry hanging on the veranda (which is about six feet by eight feet). Each depicted a baptism scene: one of St. Peter sinking into the sea and our Lord saving him; one of the Israelites in the Red Sea, when the water drowned the Egyptians; one of Naaman the Syrian bathing in the water and being cured of leprosy; and one of our Lord's baptism. The double symbolism of baptism as a saving of life and as a danger of death was shown in each mosaic. C.G was particularly impressed by that showing Peter sinking in the sea and stretching out his hand and Jesus reaching for him, this was a most beautiful mosaic of lapis lazuli. On leaving the Baptistery they went to a shop opposite to get photographs of these mosaics—one of the small shops always found near such places. They offered pictures of the Baptistery, but none of the mosaics. They went to another shop—no luck—and to several others, but they could not find the photographs they wanted. Soon after C.A. Meier was going to Italy and C.G told him to be sure to visit Ravenna and see these mosaics and get pictures*

of them, or if he couldn't to take photographs. Meanwhile G.G. was giving a seminar in the course of which he mentioned the wonderful mosaics he and Miss Wolff had seen in Ravenna, and he described them in detail. When Dr. Meier returned from Italy he told C.G. that he had gone to the Baptistery in Ravenna but that there were no mosaics there of the kind he had described. C.G. told this to Toni Wolff who said, 'That's ridiculous, I saw them with my own eyes and you talked of them for about twenty minutes!' 'Nevertheless,' he said, 'there are no such mosaics.' So at the seminar he said, 'Ladies and gentlemen, I'm sorry but there are no mosaics.'" [4-4]

Jung credited this experience for influencing development of ideas about consciousness, the unconscious, and archetypes.

"Since my experience in the baptistery in Ravenna, I know with certainty that something exterior can appear to be interior. The actual walls of the Baptistery, though they must have been seen by my physical eyes, were covered over by a vision of some altogether different sight which was as completely real as the unchanged baptismal font." [4-3]

This kind of explanation seems to posit his experience—shared by Toni Wolff—as a momentary new creation of the unconscious which arose from this thoughts about archetypal initiation, as depicted in the mosaics. Jung remembered specific details about these mosaics:

"I retained the most distinct memory of the mosaic of Peter sinking, and to this day can see every detail before my eyes: the blue of the sea, individual chips of the mosaic, the inscribed scrolls proceeding from the mouths of Peter and Christ, which I attempted to decipher." [4-3]

Jung acknowledges that viewing these mosaics with Toni Wolff was significant to him:

"The memory of those pictures is still vivid to me. The lady who had been there with me long refused to believe that what she had 'seen with her own eyes' had not existed. As we know, it is very difficult to determine whether, and to what extent, two persons simultaneously see the same thing. In this case, however, I was able to ascertain that at least the main features of which we both saw had been the same." [4-3]

If Carl Jung was alive today, and if he sent this first-hand report to me, I'd reassure him that over the past 25 years, I've received dozens of similar reports. People have noticed houses, buildings, and entire streets look completely different, as well as interior decor changing completely, such as the pig art I'd carefully examined that was completely missing on the occasion of my next visit to the Bistro Jeanty restaurant in California, apparently never having been there at all. [4-5]

ROD SERLING'S TWILIGHT ZONE: THE PARALLEL

On March 14, 1963, an episode of *The Twilight Zone* first aired on television that presented key aspects of the Mandela Effect experience, in an episode called, "The Parallel," written by Rod Serling. The protagonist of this episode is American astronaut, Major Robert Gaines, who blacks out on a mission to orbit the Earth, at the same time as ground control loses all contact with him. Despite returning safely to Earth, there are many unexplained mysteries, such as how his craft landed on land, completely undamaged, when it was supposed to make a splash-landing in the ocean. Robert Gaines is further confused when he returns home to discover that apparently, he's somehow always been a Colonel for quite some time (though he does not recall having been promoted). His house features a picket fence he does not remember, though his wife insists the house always had a picket fence. His daughter and wife behave differently with him, the president of the USA is different, and he drinks his coffee differently in what appears to be an alternate universe. [4-6]

PHILIP K DICK EXPERIENCED OTHER REALITIES

In September 1977, American author and visionary Philip K. Dick (PKD) was one of the first people to publicly discuss his experiences of different realities in a paper he planned to present as the keynote address for the Second International Science Fiction Festival in Metz, France, titled, *"If you find this world bad, you should see some of the others."* PKD wrote:

> *"I submit to you that such alterations, the creation or selection of such so-called "alternate presents," is continually taking place. The very fact that we can conceptually deal with this notion—that is, entertain it as an idea—is a first*

step in discerning such processes themselves. But I doubt if we will ever be able in any real fashion to demonstrate, to scientifically prove, that such lateral change processes do occur. Probably all we would have to go on would be vestiges of memory, fleeting impressions, dreams, nebulous intuitions that somehow things had been different in some way — and not long ago but now. We might reflexively reach for a light switch in the bathroom only to discover that it was —and always had been—in another place entirely. We might reach for the air vent in our car where there was no air vent—a reflex left over from a previous present, still active at a subcortical level. We might dream of people and places we had never seen as vividly as if we had seen them, actually known them. But we would not know what to make of this, assuming we took time to ponder it at all."

He continues:

Such an impression is a clue that at some past time point a variable was changed — reprogrammed, as it were — and that, because of this, an alternate world branched off, became actualized instead of the prior one, and that in fact, in literal fact, we are once more living this particular segment of linear time. A breaching, a tinkering, a change had been made, but not in our present — had been made in our past. Evidently such an alteration would have a peculiar effect on those persons involved; they would, so to speak, be moved back one square or several squares on the board game that constitutes our reality." [4-7]

PKD had a remarkable grasp of what the Mandela Effect is all about, from an experiencer's vantage point. Unfortunately, few people recognized the significance of what he was attempting to communicate that day in Metz, France, since,

"Just as PKD was about to start, he was informed that he had to shorten his presentation by twenty minutes. PKD quickly went through the manuscript and took out a series of sections. His translator was obliged to do the same for his document, but unfortunately, removed different sections. What resulted was utter confusion." [4-8]

Philip K. Dick lived only another four and a half years after this speech, but thanks to the existence of this written document, we can glean insights about his thoughts about and experiences with the Mandela Effect in the 1970s, long before this phenomenon was widely known.

FUTURE MEMORIES AND REALITY SHIFTS

American author and near-death experience researcher PMH Atwater coined the term "reality shifts" in her 1995 book, *Future Memory,* to describe personal Mandela Effects. Some of Atwater's reality shift reports sound just like reality shifts that have been reported in the monthly *RealityShifters* ezine, and by people sharing personal Mandela Effect experiences on blogs, social media, and YouTube. [4-9]

TELEPORTED TO SAFETY

Atwater recounts an experience shared by T.L. from Fort Worth, Texas, who was driving about 70 miles per hour in the wee hours one morning, when dozens of horses began crossing the road. All hope of somehow driving between horses was dashed, when two giant horses stood completely still in the road—so all the driver could do was brace for inevitable crash. Much to his surprise, he found himself well on the other side of the herd, driving as if nothing strange had happened at all, *"It was as if I and my car were 'transported' to the other side of the herd."* [4-10]

"Teleported to safety" turns out to be a common type of reality shift, as I've learned from documenting many such reports in the monthly *RealityShifters*, and discussing these types of experiences in talks. I was stunned the first time I mentioned people being teleported instantly to safety after giving a talk about reality shifts to see more than ten people come up to me after my presentation, to say something along the lines of, *"I haven't told anyone about this, because it sounds so impossible, but I was about to crash into an oncoming vehicle when I inexplicably ended up safe, as if we'd passed right through each other, or I was teleported."*

WHEN WE'RE NOT LOOKING, THINGS CAN CHANGE

After my kundalini awakening experience in 1994, I recognized that reality shifts must be a worldwide phenomenon affecting everyone, whether people yet realized it or not. I created a dedicated website with a simple, memorable name after being interviewed in 1998 for Elliot Stein's *Stein Online* show, and www.RealityShifters.com was born. I published the first RealityShifters newsletter in October 1999, featuring meeting Uri Geller in San Francisco. I described a "reality shift" as:

> REALITY SHIFT (ree al' i tee shift), n. 1. the manifestation of objects appearing, disappearing, transforming and transporting. 2. changes in the way we experience time. 3. any sudden, abrupt alteration of physical reality with no apparent physical cause. 4. the source of synchronicity.

Reports of items of all sizes mysteriously appearing, disappearing, and reappearing comprise a large percentage of documented cases in the first 25 years of the RealityShifters website. Everything from road signs, houses, trees, horses, people, and mountains having changed.

ALIVE AGAIN PHENOMENON

The Alive Again variety of Mandela Effect is the exact type of experience that the Mandela Effect phenomena itself is named after. In the Alive Again type of reality shift, someone is remembered to have died—and then is observed being very much alive again. The first edition of my 1999 book, *Reality Shifts: When Consciousness Changes the Physical World*, includes the first published reports of the Alive Again phenomenon with both a celebrity actor, Larry Hagman, and a pet cat having been reported dead, and later seen alive again. [4-5]

ART BELL DISCUSSES MANDELA ALIVE AGAIN IN 2001

Art Bell talked with me about what he called "the Mandela Effect," on his *Midnight in the Desert* radio show—and he was using this term about a decade before that descriptor was widely adopted for the phenomenon. Art Bell remembers it all started some time in the late 1980s or early 1990s, when someone called in to talk with him on his late night radio show, mentioning something about Nelson Mandela. Art Bell recalls responding to the caller by saying, *"Nelson Mandela died, didn't he?"* on the air, at which point Art Bell was likely the first person to publicly state that he remembered Nelson Mandela having died—long before he actually did. Art Bell said he had received *"maybe a thousand emails so far, from people who remember history a very different way,"* and *"I'm getting an awful lot of emails saying, 'Art, I remember that it happened this way—not the way it really has happened.'"* [4-11]

Throughout recorded history, we've tended to attribute Alive Again reports to merely being some kind of confusion, or mistake. Yet according to my research and surveys reported in my book, *Quantum Jumps*, 27% or 144 of the 541 survey respondents, reported that yes, they have *"seen dead people and animals alive again."* [4-12]

2005 ALIVE AGAIN SURVEY

In 2005, I conducted an "Alive Again" survey after hearing about celebrities noted "alive again" through the realityshifters discussion list at yahoo groups. In addition to mentioning actor Larry Hagman alive again in my book, *Reality Shifts*, I had heard reports that Jane Goodall had died, shortly after Dian Fossey had been murdered, in 1985. This had been greatly distressing to me, and I felt upset that people were murdering the world's great female primate researchers. Back in July 2005, the celebrity heading up the "Alive Again" survey list was Bob Keeshan, an American actor who played a TV character known as

Captain Kangaroo. A whopping 26% of those surveyed at that time recalled that Bob Keeshan had died before 2004, and then died again in January 2004. The next most-often remembered "alive again" celebrities in our 2005 survey included: Jane Goodall (15%), Larry Hagman (15%), Bob Hope (10%), Jack Palance (10%), Bea Arthur (5%), Ed Asner (5%), Walter Cronkite (5%), and Mariel Hemingway (5%). [4-13]

We might well ask, *"Why didn't anyone mention Nelson Mandela being alive again back in 2005 when that first Alive Again survey was conducted?"* which is an excellent question. At that time in 2005, neither I nor anyone I surveyed made a point to mention remembering what many of us would later recall—that Nelson Mandela had died many years earlier than his currently recorded official death on December 5, 2013.

Aside from the buzz that Art Bell noted he'd been hearing about Nelson Mandela back before April 2001, Nelson Mandela was not high on the "radar" for the realityshifters community. According to current historical reports, we see that Nelson Mandela got *huge* media coverage in February 2005, when he gave a speech in Trafalgar Square to 20,000 people. Mandela called for action to end "unnatural poverty" in a speech covered by BBC News and most major news networks. [4-14]

Some people remember seeing news that Nelson Mandela died while incarcerated on Robben Island, where he'd been confined to a small cell and conscripted to long hours of hard labor in a quarry. While the specific date of Mandela's death varies depending on the person remembering it, there are many similar details recalled. For example, many people have vivid memories involving the South African government negotiating Mandela's estate with Mandela's widow, Winnie, for example.

Current historical records show that Nelson Mandela died at home in Johannesburg on December 5, 2013, after battling a respiratory infection. South African citizens flocked to his home to pay tribute after his death, which was mourned by people around the world.

MANDELA EFFECT-AFFECTED TOWN

While investigating early reported incidences of the Mandela Effect, I discovered that residents of a town in the United Kingdom were concerned that their beloved dinosaur had gone missing. A full-scale model of a dinosaur was an oft-mentioned favorite memory from the 1960s for many visitors to the Bolton Museum and Art Gallery, as described in a January 2006 article in *The Bolton News* —yet there were no official records that any such dinosaur existed. This was a matter of importance to the community, since restoring the old exhibit would have saved the museum time, effort and energy. Alan Rushton, leader of the Conservative group on Bolton Council, said:

> *"One of our dinosaurs is definitely missing! Am I supposed to accept that my eyes deceived me, or this enormous dinosaur was a figment of my imagination? I remember as a schoolboy going on visits to the museum and standing in awe, looking at it. I raised the matter when I was on the arts committee, but we were never able to find any trace of it."*

And Anne Wright, head of Rumworth School in Bolton, England, also remembered seeing the dinosaur when visiting the museum:

> *"I remember a huge dinosaur standing in the entrance to the natural history section. One day I went in and it was no longer there. Later on, when I asked about it, none of the staff could remember anything about it, but I know I saw it. Now nobody seems to know anything about—it's bizarre."*

About half of those responding to the Bolton News story remembered the Bolton dinosaur with no official records of it existing there. [4-15]

FIONA BROOME POPULARIZES "MANDELA EFFECT"

In 2010 while Nelson Mandela was still alive, Fiona Broome famously popularized the term "Mandela Effect" after talking with people who remembered Nelson Mandela having passed away years earlier. While Art Bell was the person I first heard discussing the Mandela Effect, the term did not grab public attention until Broome helped to popularize it.

Google trend reports show the term Mandela Effect breaking through to public awareness in April 2016, where it maintained a steady presence in popular culture ever since. The Mandela Effect has been covered by numerous media outlets, including: *CNN, Forbes, Parade, Psychology Today, USA Today, Popular Mechanics, Readers Digest, Seventeen, The New Zealand Herald, The Globe and Mail, San Diego City Beat, Discover Magazine, The Atlantic, Cosmos, Vice,* and CNBC. Mandela Effects most often mentioned in media reports include: The Berenstein Bears, Haas avocados, the lion laying down with the lamb in the Bible, Stouffer's Stove Top Stuffing, the color of C-3PO's legs, Curious George having a tail, the hyphen in Kit-Kat, the location of the kidneys and heart in the human body, Jiffy peanut butter, and the witch in *Sleeping Beauty* saying, *"Mirror mirror on the wall."*

MANDELA EFFECT AUTHORS AND THEIR BOOKS

In 2006, psi researcher **Starfire Tor** published some of the first reports of time shifts, reality shifts, and the Mandela Effect on her website www.starfiretor.com. In May 2006, Whitley and Ann Strieber interviewed her on the *Dreamland* and *Unknown Country* radio shows. Starfire Tor was featured in *This Book is from the Future*, talking about time shifts, the core matrix, co-existing time lines, reality shifts, time travel, and the impact of these on humanity and the Earth. [4-16] [4-17]

In 2012, author **Trish LeSage** published, *Traveling to Parallel Universes,* describing physical changes she observed in the days and weeks prior to switching between universes—including changes in her body temperature. LeSage noticed differences in the way things manifested, and changes in people and animals, alerting LeSage when she arrived in each parallel world. Trish noticed the presence and/or absence as well as changes in personalities of the local wildlife and stray cats she cared for. Her open-mindedness, pure-heartedness, and focus on the here-and-now was an asset when finding herself in worlds where things changed so dramatically. LeSage noticed some celebrities who had been alive sometimes had now long been dead, familiar TV shows were sometimes advertised as if starting for the first time, and personalities of people and animals were entirely different. In one intriguing account, Trish described a universe in which people could easily manifest whatever they contemplate relatively quickly—yet due to lack of character-building through hardships, people there were not very friendly, genuine, nor heart-centered. [4-18]

Australian psychology professor **Dr. Tony Jinks** created a typology for personal Mandela Effects and reality shifts in his 2016 book, *Disappearing Object Phenomenon (DOP)*, based on 385 case studies. Dr. Jinks confirms that experiencers are just as mentally competent as anyone else, by ruling out that experiencers of Disappearing Object Phenomenon are psychologically distinguishable from non-experiencers. Jinks demonstrated that experiencers are not suffering from such possible psychological issues as: transience, telescoping, errors of commission (i.e. suggestibility, illusionary correlation, and confirmation bias), perceptual blindness (i.e. inattentional blindness), altered states of consciousness (i.e. hallucinations and auto-hypnosis), encoding errors (i.e. absent mindedness), memory distortion (i.e. pre-existing beliefs), and loose associationism. Dr. Jinks' typology of DOP includes: Disappearances, Appearances, Reappearances, and Replacements. Jinks notes that DOP experiences are typically isolated events occurring in the form of discontinuities, with a seemingly intact historical sequence of prior events supporting the current status quo. Jinks contemplates possible causes of DOP such as: invisibility, teleportation and/or wormholes, parallel universes, dimensional shifts, external agents, and human-centered causes, unwittingly caused by the experiencers themselves, for various obscure subconscious reasons. Because these DOP incidents are characterized by a quality of seamlessness, Jinks favors the explanatory theory of the idea of conscious observation (by multiple consciousness states) responsible for "collapsing reality," so we have our familiar consensual reality, such that the new situation appears as if it has always been that way. [4-19]

Spiritual teacher and Biblical scholar **Stasha Eriksen** shared first-hand reports of Mandela Effects in her 2017 book, *The Mandela Effect: Everything is Changing*. Stasha provides rich descriptions of Mandela Effect experiences that she and others have encountered, and includes chapters covering changes to: human anatomy, astronomy, and the Bible. Eriksen also presents theories as to what, exactly might be causing the Mandela Effect. [4-20]

In 2018, **Anthony Santosusso** published an autobiographical account of his Mandela Effect experiences in *Mind Beyond Matter: The Mandela Effect*. One pivotal moment in Santosusso's Mandela Effect odyssey involves miraculous survival of what he expected to be his impalement. Instead of finding himself skewered by a sharp object, everything was strangely fine. This incident led Santosusso to consider the notion that all people who see the Mandela Effect die in some reality—yet have experiences similar to his, of miraculously surviving. Santosusso states, *"If there is one thing I would like for you to understand, it is that these things aren't changing. The song lyrics did not change, you did! Nothing is changing, except for your consciousness and perspective."* I respect Santosusso's invitation to people to contemplate the Mandela Effect with their own thoughts, information, and perspective. [4-21]

In August 2018, **Rob Shelsky** published, *Shattered Reality: The Mandela Effect,* pointing out that many people who've experienced the Mandela Effect reject the popular definition of "false memories," in favor of considering the phenomenon to be "mass memory discrepancy." Shelsky prefers this term, *"because it specifically refers to 'mass' as in masses of people. The term also refers to 'discrepancy,' suggesting an inconsistency, incongruity, or divergence."* While acknowledging that false memories can happen, he points out three major reasons that the Mandela Effect deserves special attention: (1) the number of people experiencing the Mandela Effect can be huge, (2) specific memory discrepancies tend to match for large groups of people, and (3) there exists actual physical evidence in some cases, indicating that what people remember may actually have been true. Shelsky shares evidence for such things as remembering how the human heart used to be on the left side of our bodies (not the center), as shown in the *Boy Scout Handbook* that specifically mentions a left-handed Boy Scout handshake, *"made with the hand nearest your heart."* Shelsky notes that many Mandela Effect changes "have been in the form of improvements to human anatomy, almost as if someone was altering us in some way to make us better, at least physically," noting changes to: eye sockets, kidneys, and the nasal cavity. Shelsky observes that we no longer have floating ribs, so organs are better protected, and it's no longer so easy to slash one's wrists, since the arteries have moved closer to the bone. He points out, "The changes are all making us stronger, more protected, and better able to function." Shelsky includes fascinating photos from movies and TV showing now-missing land masses, such as a rather large island off the southwest coast of Australia that is not Tasmania appearing in the movie, *Dazed and Confused,* and the TV show, *Friends.* Phantom islands come and go, but for whatever reason seem more likely to vanish than appear, such as Antilla, Sarah Ann Island, Hy-Brasil, Buss Island, and others. [4-22]

JOTT stands for "Just One of Those Things," and involves the way that seemingly everyday objects can move almost as if with their own accord. British barrister, parapsychologist, and president of Oxford University Society for Psychical Research, **Mary Rose Barrington,** affectionately called such incidences "jotts" or "jottles," and sorted them into various typologies. "Flyaways" are objects that vanish, "Turn-Ups" are objects that appear, "Windfalls" are objects that show up that had never previously been around, "Comebacks" are objects that disappear and then reappear, "Walkabouts" refer to objects that seem to move to some new location, and "Trade-Ins" are items that seem quite different than what they used to be. I love how JOTT was written by a researcher born in the 1920s who was keenly aware that this JOTT phenomenon was a matter of common knowledge over a century ago. Some of the detailed reports included in JOTT occurred as early as the 1950s, and Barrington points out that the history of the phenomenon goes even farther back. Barrington writes:

> *"In our time, we seldom hear reports of sensational macro-phenomena said to have taken place during the 100 years starting mid-nineteenth century, when many people believed that such things were possible and that they could be evoked by the human endeavor, usually with input from the supposed spirit world."*

JOTT's first-hand reports came from submissions following an article from the magazine, *The Unexplained,* which asked readers to write in with their experiences. Barrington considers the JOTT phenomenon to be related to Cosmic Mind, since these experiences are relatively widespread, and have been around for quite a long time. I am deeply grateful to have found Barrington's informative and engaging research, after having dedicated more than twenty years to studying this exact same phenomena, and never having heard of Barrington, her work, the terms "jott" and "jottles," or her book until after she died. [4-23]

We can gain insights to the reality shift and Mandela Effect phenomenon by reviewing the types—or typographies—independently developed by researchers Mary Rose Barrington, Tony Jinks, and myself, as shown in Table 4-1. Mary Rose Barrington noted six types of JOTTs: Fly Aways, Turn Ups, Windfalls, Comebacks, Walkabouts, and Trade-Ins. Tony Jinks documented three types of DOPs: Disappearances, Appearances, and Reappearances. I noted seven types of Reality Shifts: Disappearances, Appearances, Reappearances, Transportations, Transformations, Time Shifts, and Alive Again.

Table 4-1:

JOTT, DOP, and Reality Shifts Typographies

Mary Rose Barrington *JOTT*	Tony Jinks *DOP*	Cynthia Sue Larson *Reality Shifts*
Fly Aways	Disappearances	Disappearances
Turn Ups	Appearances	Appearances
Windfalls	Appearances	Appearances
Comebacks	Reappearances	Reappearances
Walkabouts	Reappearances	Transportations
Trade-Ins		Transformations
		Time Shifts
		Alive Again

In 2018, **Bill Bean** published *Stranger than Fiction: True Supernatural Encounters of a Spiritual Warrior,* including a chapter dedicated to the Mandela Effect. Bean recalls that the family of bears featured in popular children's books were known as the "Berenstein Bears," not the "Berenstain Bears." Bean first noticed the Mandela Effect in 2015, and gave the topic more thought when he observed changes to the Bible. One striking Bible change involves a well-known passage from Isaiah 11:6, which Bean notes, *"was one of the most well-known scriptures in the entire Bible. There are statues, paintings, songs, movie lines, figurines among other things based on this scripture."* That famous biblical scripture now reads, and has always read, *"The wolf also shall dwell with the lamb, and the leopard shall lie down with the kid, and the calf and the young lion and the fatling together, and a little child shall lead them."* [4-24]

In 2018, **Dale DuFay** published *Terra's of the Milky Way,* presenting the premise that there may be countless numbers of ourselves, on many possible Earths. DuFay observed older Earths on the outskirts of the galaxy, with younger Earths closer to galactic center. DuFay emphasizes that the Mandela Effect is a gift from Jehovah stating, *"I'm certain that once everyone knows they have a chance to write the final chapter in their very important book of life, they'll be more apt to be the best human that they can possibly be!"* DuFay suggests we live in a parallel existence, reborn into mortal, imperfect human short-lived avatars, on our way to becoming immortal avatars. DuFay reports personal changes including: changes in his facial scars, and changes to his pets' appearances and personalities. DuFay also details differences between various Earths and events in World War II, and differences in the way dinosaur bones have been displayed on different Earths. DuFay's central premise is that the Mandela Effect is a holy event prophesied in the Bible, and hinted at in some parables, in the book of Revelation, and in the gospels of Jesus. DuFay adds, *"We can't be discouraged by the fact that the Bible is different on some Earths compared to other Earths, in certain texts. The very essence of the Bible remains, and that's the most important thing here."* [4-25]

In April 2019, **Tray S. Caladan** published *Mandela Effect: Analysis of a Worldwide Phenomenon.* Caladan provides some of the most comprehensive written documentation of Mandela Effects, including numerous inexplicable changes that have occurred in so many areas, affecting statues, planets, pyramids, works of art, movies, songs, logos, our bodies, the Bible and more, indicating that something bigger than simple mis-remembering is at work. Caladan asks us to consider whether the large pyramids at Giza have switched positions, and if the Moon and Mars seem different. Have some countries changed borders and sizes, as well as entire continents shifting, seemingly overnight?

Caladan dismisses the notion that the Mandela Effect is merely a matter of mistaken memories, and suggests we've been pushed or forced into a darker, parallel world. [4-26] [4-27]

In 2019, YouTubers **Eileen Colts, Paulo M. Pinto, Shane C. Robinson,** and **Vannessa VA** co-authored, *Mandela Effect: Friend or Foe?* Big, inexplicable, instantaneous changes are being observed in movies, books, art, logos, geography, human anatomy, celestial constellations—and every conceivable type of historical fact and event. Their book is an excellent resource for understanding, coming to terms with, and ultimately embracing the staggering implications of the Mandela Effect. Through sharing strengths of their individual viewpoints, perspectives and insights on this multifaceted topic, the four co-authors present a big picture of what they observe, and what each of them suspects might be going on. [4-28]

In October 2021, **Rizwan Virk** published *The Simulated Multiverse: An MIT Computer Scientist Explores Parallel Universes, the Simulation Hypothesis, Quantum Computing, and the Mandela Effect.* Virk brings a refreshing scientific and computer science perspective to what's going on with the Mandela Effect. He also presents some fascinating insights about such things as what Philip K. Dick had in mind when writing his novel, *The Man in the High Castle,* that Virk discovered when talking with PKD's wife, Tessa, and listening to PKD's Metz speech where PKD said, *"We are living in a computer-programmed reality, and the only clue we have to it is when some variable is changed, and some alteration in our reality occurs."* Virk's epiphany was that PKD was not only stating that we are living in a simulated reality, but also that multiple timelines are running. Virk's book makes a brilliant case for why a simulated multiverse might be the best model for understanding reality. [4-29]

In 2023, **Anthony J. Duran** published *Breaking Reality: Inside the Mandela Effect,* sharing some commonly noted Mandela Effect examples, and stating that he feels drawn to the idea of collective consciousness as a possible explanation for the Mandela Effect. [4-30]

MANDELA EFFECT EVENTS

The first official in-person Mandela Effect event was held in the USA in a meet-up at a restaurant in **June 2017 in Dallas Fort Worth, Texas**, arranged by YouTuber Guy Fauqes. A group of eight Mandela Effect experiencers gathered together, including YouTubers Guy Fauqes, Scarab Performance, and Harmony Mandela Effect. While this event featured no audio or video recording, YouTuber Harmony

Mandela Effect created a video shortly afterward, describing some of what he most appreciated about it. This Mandela Effect event was significant for being the first official Mandela Effect gathering of any kind. [4-31]

Author and YouTuber Vannessa VA organized a Mandela Effect conference in **July 2018 in Manassas, Virginia**. An official announcement pointed out that the main reason for the gathering is that, *"We are scattered sparsely among the sleeping."* The invitational video described how there are three groups of people: those who don't see any changes, those who see changes but don't care that reality itself is changing, and those who see changes and appreciate the enormity of the implications of the Mandela Effect. Participants at this 2018 conference included YouTubers: Meegs B, SMQ A.I., John Boyle, and Dewayne Hughes. Topics included how some of the Mandela Effect Bible changes appear to be a message from God, telling us what is going on, to those of us who have eyes to see: *"the wolf is dwelling with the lamb, and that is the time that we are in."* [4-32] [4-33]

2019 IDAHO CONFERENCE

Jerry "DarkWolf" Hicks organized the first International Mandela Effect Conference (IMEC) in **November 2019 in Ketchum, Idaho**. All talks were recorded with video recordings posted on the International Mandela Effect Conference YouTube channel. [4-26] Participants in this conference included: author Shane C. Robinson from the YouTube channel Unbiased and On the Fence, Quantum Businessman Christopher Anatra, Canadian artist Kimberly-Lynn Hanson, and myself. The mission of this event was to create a Mandela Effect conference experience that was fun, informative, and inspiring. People described this conference as feeling like a magical, surreal, and amazing retreat. I loved being able to say, *"as I remember, it used to be different,"* and see nods of understanding. I also appreciated how this event was created with a feeling of cooperation toward a common goal, free from ego, and filled with respect.

Many attendees with YouTube channels enjoyed informal conversations and discussions at this November 2019 Idaho event, including: Eva from Once Upon a Timeline, author Anthony Santosusso, Guy Fauqes, and JW. A key theme emphasized at this event by speakers Chris Anatra, Shane Robinson, and Kimberly-Lynn Hanson was that when we look for it, we can find deeper truth and meaning evident in Mandela Effects, and the perspective we choose to adopt influences what we actually see. Kimberly-Lynn presented several examples of Mandela Effect art that invite us to answer the question for each piece, *"How is this closer to truth?"* Such a line of inquiry reveals that the change from "Jiffy" peanut butter to "Jif" invites us to contemplate, *"What could possibly be jiffier than Jif?"* Kimberly pointed out that spelling the bears name as "Berenstain" means reduced pronunciation or spelling confusion—there will be no child left behind! Kimberly discussed some first principles for living a nonlinear lifestyle, which include: What's wrong is right, Nature makes no mistakes, and Put your eggs (intentions) in one basket. The principle about putting your eggs in one basket reminds me of quantum coherence, and the idea I share in my book, *Reality Shifts*, about minimizing destructive interference of subconscious thoughts. All too often, people sabotage hopes and dreams with negative thoughts and feelings about who they feel they are, what they think they should be doing, and how they believe things ought to be different.

A conference attendee from Sweden corroborated geographical changes presented by Shane Robinson, who showed how he had painstakingly placed the word "love" in many languages across all geographic regions on a world map—and Shane could clearly see how the world map had changed. The Swedish woman recalled she had frequently traveled to Denmark, which seems now to have moved. Shane's map had no words for "love" where Denmark is now, and she said she remembered traveling by ferry boat down to where Denmark used to be. Shane pointed out that we can see a truth in Rodin's statue, *The Thinker,* changing from elbow on right to left side reflecting how our modern culture now thinks predominantly in a left-brain, analytical mode—rather than with a more intuitive right-brain approach. [4-34]

Chris Anatra discussed the meaning of changes in eye color and hat color of actors who starred on the *Gilligan's Island* TV show. Anatra shared a videotaped interview with his parents, who noticed changes in songs the family sang together, and changes in the Bible. Anatra shared changes he noticed in his business, with clients and data, and pointed out changes noted in works of art and historical events. [4-35]

In my Friday presentation at the Idaho conference, I mentioned how I had experienced bi-location. [4-36] Bilocation was subsequently experienced by conference attendees Moriah and her son, two days later on November 10th. That evening, Moriah commented to a Canadian participant how much she enjoyed chatting with her for about 20 minutes in front of the hotel earlier that day. The Canadian participant looked confused, and said aside from breakfast that morning, she had spent the rest of the day in her room. I asked her if she'd been daydreaming—and she replied, yes, she had been thinking about the very same subjects that Moriah recalled having discussed with her about how some people don't grasp the Mandela Effect, and how to productively address negativity in non-combative ways.

Our 2019 Idaho conference also featured a couple of cases of people popping into our reality as if they'd always been there. I discussed with Cat and Kimmie (the Medium who Rocks) that Dolores Cannon was new to me. We recalled an author who was similar to Dolores who we remembered, Michael Newton, author of *Journey of Souls* and books similar to Cannon's, so surely we'd have seen her books side-by-side with his. When I look up Newton's book, *Journey of Souls* now, I see the first edition was published in July 1994. Cannon's, *Keepers of the Garden*, states it was written in 1993—yet I'd not seen it in the 1990s.

After the Idaho conference, I chatted with Joe Rupe on his *Lighting the Void* radio show on November 25, 2019. Joe raised the subject of how much he'd enjoyed reading the book *Journey of Souls* by Michael Newton, so I asked if he recalled seeing Dolores Cannon's books decades ago, when he was reading Newton's books. Rupe replied that not only did he not remember reading any of Cannon's books, but by all rights he should have been aware of Dolores Cannon, since her publishing company was Ozark Mountain Publishing, "just up the road" from where Rupe lives in Arkansas! Joe and I only recalled first having seen Dolores Cannon pop up in our subjective experience around 2017.

After the Idaho conference, on November 13, 2019, Eva from the YouTube channel "Once Upon a Timeline" noted in a Moneybags73 chat (about Silence of the Lambs), *"Cynthia is an ME for me, did not exist until recently, it was just Fiona,"* recalling Fiona Broome, but not me, until we met in Ketchum, Idaho.

2019 ST. LOUIS MIDWESTERN CONFERENCE

Speakers at the Midwestern Mandela Effect Conference (MMEC) held **November 15, 2019 in St. Louis, Missouri** included: Dr. Sharon Squires and author Bill Bean. Sociology professor Dr. Sharon Squires presented examples of some noteworthy Mandela Effects including: human physiology (changes in location of the heart and kidneys), geographic changes (disappearance of the north pole), and technology that existed decades earlier than most of us recall, such as 1940s Panoram video jukeboxes. Squires pointed out how concerning many changes are to people, since they affect works of art and sacred religious texts, including the Bible. While memory problems might be a possible cause for the Mandela Effect, that hardly explains the uniformity of false memories, nor the diversity of experiencers. Squires considered eight possible explanations for the Mandela Effect, including: psychological problems, social hysteria based on some trigger event, religious causes, simulation theory, CERN merging realities together, Natural causes, we have all died, and nefarious external technology. [4-37] Reverend Bill Bean addressed concerns about Bible changes in his talk about Bible changes, saying he is not trying to scare or manipulate anyone, despite stating he believes the Mandela Effect, "is the work of the devil." [4-38]

2020 IMEC EVENT

The International Mandela Effect Conference (IMEC) became a 501(c)3 nonprofit organization in 2020, and held its first official event as a free, online livestreaming conference on YouTube in June 2020. Jerry Hicks ("DarkWolf") was the Master of Ceremonies for this event, with a keynote by Regina Meredith, and presentations by fourteen speakers. Speakers at IMEC 2020 included: Christopher Anatra, Sean Bond, Eileen Colts, Nicole DeMario, Jan Engels-Smith, Kimberly-Lynn Hanson, Jerry Hicks, Cynthia Sue Larson, Akronos Mago, Evan Matraia, Eva Nie, Shane Robinson, and Anthony Santosusso. This event included a "Mandela Effect community montage" featuring 10 second clips from 34 YouTube channels. Shane Robinson noted that a Mandela Effect occurred during his conference in his IMEC 2020 presentation: there had been further changes in the locations of land masses on his "love" map since the November 2019 conference.

2023 IMEC EVENT

The International Mandela Effect Conference resumed live gatherings in **September 2023 in Branford, Connecticut**, with a conference organized by Jerry "DarkWolf" Hicks acting as Master of Ceremonies, Christopher Anatra, Shane Robinson, and myself. Speakers at IMEC 2023 included: Tom Campbell, Roger Marsh, Christopher Anatra, Sean Bond, John Kirwin, Meegs B, Shane Robinson, Dan Hennessey, Tete DesJours, Jerry Hicks, Alicia Thompson, and myself.

MANDELA EFFECT ON RADIO, TV, AND FILM

As the number of Mandela Effect experiencers rise, so too does media coverage of the Mandela Effect and Mandela Effect experiencers. The *X-Files* TV show is famous for exploring "fringe" topics, such as UFOs, extra-terrestrials, time travel, aliens—and also, the Mandela Effect. *The Lost Art of Forehead Sweat* episode of the *X-Files* aired in 2018, treating the topic in a light-hearted humorous manner.

In June 2019, the *New York Times* Sunday crossword puzzle featured the Mandela Effect.

In 2020, the TV show *Jeopardy* added "The Mandela Effect" category, including this question: *"Despite what you might remember, this actor is not wearing sunglasses when he dances to Old Time Rock & Roll in Risky Business."* *Jeopardy* questions and answers are crowd sourced, so the fact that the Mandela Effect gained its own category on this show is a strong indication that the Mandela Effect is going mainstream. [4-39]

In 2020, AJ "the Ripon Rabbit," and Evan Matraia (aka moneybags73) hosted the *Mandela Monthly* radio show on YouTube, featuring call-in questions and conversation with listeners in the chat room during livestreams. [4-40]

While Mandela Effected experiencers may not yet feel like they are taken seriously or treated respectfully, as the Idaho Mandela Effect conference organizer Jerry Hicks pointed out, the Mandela Effect topic is at a similar point now to where the UFO community was just a couple of decades ago. Those who had been mocked for having claimed to have seen UFOs are now being vindicated, as the United States government admitted in December 2017 the existence of government Unidentified Flying Object (UFO) programs, also known as Unidentified Aerial Phenomena (UAP). While former Pentagon officials don't speak on behalf of the government, the presence of several such former high-level officials lends credence to those who have reported their UFO experiences. [4-41] In May 2023, a survey of 1,460 scholars from 144 universities showed that 19% of those professors, scientists and scholars reported having seen UFOs, also known as UAPs. [4-42]

In December 2019, *The Mandela Effect* movie invited us to enter the world of Brendan, a father dealing with devastating personal loss, as he discovers that reality is changing. Brendan observes changes to games, cartoons, children's book titles and characters, peanut butter, and more. Though Brendan attempts to share what he learns about the Mandela Effect with his wife and brother-in-law, they don't share his zeal to discover what is going on. *The Mandela Effect* presents several possible theories for what might be causing people to collectively mis-remember so many things, and zeroes in on one particular theory that resonates for Brendan. *The Mandela Effect* is a good conversation-starter for those who are either new to or well familiar with the topic, providing insights as to how it feels to be living adjacent to similar, yet slightly different possible realities. [4-43]

I had the opportunity to ask the movie's director, David Guy Levy, a question in December 2019 after his movie had just come out, "*Do you have advice for people who realize this might be a real phenomena, this Mandela Effect?*" David replied,

> "*I would say... I was looking for advice myself probably, in the process of making this film—and working through all these ideas. And the one question that I kept asking myself was: if our reality is not the way we believe it to be, and if there's something going on here, does that mean we should let it unsettle us? Or should we let it make us question if anything is really worthwhile? And I think my advice then would be to ask themselves... ... let's say the answer for what's causing it is what I'd gravitate to. And I know everyone has their own opinions. People believe in a few different theories for what could explain this effect. And the one I gravitate towards is simulation theory—the simulation hypothesis. There's a lot in it I just gravitate to... ... And then the question that came with that is, is any of this meaningful? What's the point? How do you not become a nihilist, if that's what you believe? And the advice I have is that, don't forget that you're a feeling being. Because, even if you have these memories of things in your life—your personal life—the fact that you get to feel these things in the first place should be real. And that's the line that we put that in the movie, because the main character loses his daughter, and all the memories he has of her are changing. He asks, 'Is any of this real? I want the truth—I want to know the truth!' and the advice he gets is, 'If you remember her and the smile she put on your face; if you remember the tears that rolled down your cheek; you get to feel those things; it should be real enough.' And I think not dismissing the beauty that we still have in our lives, and even though reality might not be what we hoped it might be, or things are just out of our grasp, because we're still animals, and we're still not going to know the truth of the universe as much as we might want to, my advice is: don't let this depress you.*" [4-44]

I recommend adopting an optimistic perspective when noticing the Mandela Effect and reality shifts, since with gratitude, we can have hope, and with forgiveness we can experience wisdom. I have witnessed substantial differences between subsequent shifts when feeling caught in an emotional downward spiral, or in an energized relaxed, detached state of mind.

An episode of an HBO docu-comedy series of *How To... with John Wilson*, "How to Improve Your Memory" was released in 2020. The description for this show is, *"An anxious New Yorker who attempts to give everyday advice while dealing with his own personal issues."* This episode includes footage from presentations and interviews with participants at the 2019 Idaho conference, as well as some sections filmed with Quantum Businessman, Christopher Anatra, in Connecticut. John Wilson appears to have experienced the Mandela Effect, as evidenced by the last few minutes of this episode, in which a song by Meatloaf, "Objects in the Rear View Mirror May Appear Closer than they Are," plays while Wilson modifies some common products on grocery store shelves, putting them back the way so many of us remember. [4-45]

An episode of *Robot Chicken* in 2022 playfully included the ghost of Nelson Mandela as a character providing time travel and timeline correction possibilities to a well-intentioned nerdy character. The nerdy character who attempts to go back in time and "correct" some well-known Mandela Effects obtains some humorous results. [4-46]

MANDELA EFFECT RESEARCH

In 2022, many publications covered news of a University of Chicago study finding strong evidence that the Mandela Effect is real, and that the Mandela Effect is widespread and consistent. [4-47] [4-48] Wilma Bainbridge and Deepasri Prasad studied why people remember certain things over others, and were surprised to discover evidence that people tend to share extremely similar "false" memories of 40 different icons in the study. [4-49] Amazingly, large numbers of people shared similar specific memories of logos and characters, describing, for example, Monopoly's Rich Uncle Pennybags or Pokémon's Pikachu with identical "incorrect" details. [4-50] While including such biased descriptive wording as, "false memories," their conclusion is significant: *"We revealed a set of images that cause consistent and shared false memories across people, spurring new questions on the nature of false memories."* [4-48]

* * * * * * *

The next chapter, *Mandela Effect Theories,* explores some of the most intriguing Mandela Effect theories, and invites consideration and discussion about what might be causing this remarkable phenomenon.

~~~~~~

EXERCISE:
MANDELA EFFECT TV SHOW SAFARI

Some of us have noticed many changes to dialogue, eye color, costumes, famous catch-phrases and more in popular TV shows. We remember TV characters saying things they've now supposedly never said, such as *Star Trek's* Captain Kirk saying, *"Beam me up, Scotty."* Watch some of your favorite TV shows, with an eye to noticing anything that seems different than how you remember it having been the last time you saw the show.

~~~~~~

EXERCISE:
TRACKING JOTTS, DOPS, AND REALITY SHIFTS

Considering the different ways things have long been noted to disappear (fly away), appear (turn up and windfalls), reappear (come backs), and transport (walkabouts)—as well as transformations (trade-ins) and time shifts, with occasional instances of people showing up or being alive again—which types of personal Mandela Effects do you most enjoy, and tend to notice most often?

~~~~~~

"*I regard consciousness as fundamental.
I regard matter as derivative from consciousness.
We cannot get behind consciousness.
Everything that we talk about,
everything that we regard as existing,
postulates consciousness.*"

—Max Planck

Chapter 5

MANDELA EFFECT THEORIES

Most everyone who acknowledges the reality of the Mandela Effect immediately wonders how and why it is happening. There are dozens of possible explanations. Some of the most plausible theories include:

False memories and confabulation

Simulation Theory / Dream

Scientists at CERN

Quantum phenomena

Quantum immortality

Time travelers

Converging timelines / Reality bubbles

Nonlinear, orthogonal time shifts

Spiritual apocalypse

Schumann resonance

Evolution of consciousness

These and other Mandela Effect theories can be combined, with numerous variations. Some popular combinations include: Quantum immortality combined with spiritual apocalypse, simulation theory combined with CERN scientists running amok, and quantum phenomena combined with evolution of human consciousness. In this chapter, we'll take a look at the above Mandela Effect theories, with the exception of quantum phenomena, which will be covered in detail in Chapter Six, *Science of the Mandela Effect*.

FALSE MEMORIES AND CONFABULATION

In 2009, psychology researchers in Italy and the U.K. published a paper, "Collective representations elicit widespread individual false memories," that stated, *"It is well known that the need to simplify a story, to preserve its consistency and to avoid discontinuity gaps, leads to memory distortions. These biases allow one to preclude possible dissonances between the accurate recollection of past events and one's own current knowledge or present situations upon which one should act."* This paper mentions the fascinating case of the clock at the Bologna railway station, which was left stopped at 10:25, marking the time of a terrorist event. Researchers believe that this appears to be a case of local Italian people having a "distorted memory" of their train station clock being fixed at the time of the explosion, never operating in the intervening years. The clock stoppage conferred an iconic significance, with many locals recalling that the clock had been stopped ever since that terrible event. Historical records show that the clock had only been stopped much later after that event. The researchers present the theory that because memory is reconstructive, we can therefore expect people to sometimes revise memories in similar fashion with one another for events of collective social meaning and significance. [5-1]

The conflation between false memory and Mandela Effect has become so strong in mainstream posts that Wikipedia mentions the Mandela Effect as being an example of false memory:

> *In 2010 this phenomenon of collective false memory was dubbed the "Mandela Effect" by self-described "paranormal consultant" Fiona Broome, in reference to a false memory she reports, of the death of South African leader Nelson Mandela in the 1980s (rather than in 2013 when he actually died), which she claims is shared by "perhaps thousands" of other people. Other such examples include memories of the Berenstain Bears' name previously being spelled as Berenstein, and of a 1990s movie, Shazaam, starring comedian Sinbad as a genie. [5-2]*

Associating false memory with the Mandela Effect encourages people to jump to the conclusion that the Mandela Effect shares similar causes as previously recognized types of false memories—such as the Construction hypothesis for malleability of memory, or Skeleton theory, in which a memory is recalled in two categories (acquisition and retrieval) consisting of several steps where memory errors can be introduced in memory construction. People may feel encouraged to assume that these theories can completely explain away the Mandela Effect, because some experiments demonstrate that imagining events can sometimes lead people to subsequently develop false memories.

Some memory experts point out that people naturally have "defective" or false memories, since people are suggestible, and this combines with social tendencies toward confabulation. Some memory experts point out that we naturally alter our memories each time we retrieve them.

Studies by psychologists such as Dr. Elizabeth Loftus suggest that the way questions are framed following an event can influence how people remember the sequence of events, illustrating just how malleable human memories can be. In an experiment conducted in 1978, Loftus and her colleagues showed study participants several images of a car at a junction, and later questioned the participants about the scene they had observed. Some study participants were asked whether they had seen "a stop sign," while others were asked if they had seen "the stop sign." Those participants who were prompted with the word "the" were more likely to recollect it than the other group, with a very small change in wording seeming to assure people that an object exists. [5-3]

In July 2020, memory expert Dr. Elizabeth Loftus accepted an invitation to discuss the Mandela Effect on an episode of the *Mandela Monthly* radio show featuring AJ, "the Ripon Rabbit," and Evan Matraia (moneybags73). Dr. Loftus indicated her evident bias regarding the cause of the Mandela Effect throughout this conversation, by variously referring to it as: "memory distortion" and "mis-remembering." By the end of this conversation with professional stage magician A.J. and Mandela Effect researcher, Evan Matraia, Dr. Loftus acknowledged that she remembered two things rather differently than has apparently "always been true," and expressed openness to learning more about the Mandela Effect. Dr. Loftus remembered the (currently non-existent) line of dialogue in The *Wizard of Oz* movie, *"Fly, my pretties, fly!"* Dr. Loftus also remembered another currently non-existent closing line at the end of every George Burns and Gracie Allen TV show where George would say to Gracie, *"Say good night, Gracie,"* and Gracie would say, *"Good night, Gracie!"* [5-4]

While acknowledging fallibility of human memory, we can benefit from further investigating—rather than dismissing—the relationship between the Mandela Effect and memory. Why do so many people agree upon specific, identical details about things that never were? Intriguingly, some research studies involving "flashbulb" memories may provide us with clues to how we may be living in parallel worlds. Research results indicate students didn't always recognize their own hand-written notes of where they were, who they were with, and what they were doing at the time of the Challenger space shuttle explosion on January 28, 1986. Upon seeing their hand-written notes one year later, some said,

"That's my hand writing, but that's not what happened." [5-5]

Some suggest that the implication that Mandela Effects are "false" memories is based on the assumption that recorded events are automatically assumed to be "true," so therefore people must be mistaken in thinking that they remember specific things differently from recorded historical facts. A presumption to call Mandela Effects "false" memories assumes a bias in favor of a classical physics view of reality without acknowledgment of properties of quantum physics appearing in macroscopic systems and our daily lives.

Proponents of the notion that all Mandela Effects are false memories succumb to the logical fallacy of hasty generalization. As Walton writes in "Argumentation Schemes for presumptive reasoning":

> *"The fallacy of hasty generalization needs to be understood not simply as a flaw or error in one type of reasoning, but as a kind of scalar fallacy where reasoning is pushed forward from one of five levels to another."* [5-6]

We can thus witness a perfectly reasonable statement at level one (Argument from Example): "This Mandela Effect is a false memory." A person might have some reasonable basis for such an assertion. If someone takes this to level two (Particular Claim) we get, "Some Mandela Effects are false memories," which might be true. At level three (Plausible Generalization) we arrive at, "Typically, Mandela Effects are false memories," which is quite a logical leap. At level four (Inductive Generalization), someone might say, "Most Mandela Effects are false memories." By level five (Universal Generalization), the bold assertion might be made that, "All Mandela Effects are false memories."

While considering the reliability of our minds and memories, we do well to take a look at studies conducted by Australian neuroscience professor Dr. Tony Jinks. These studies help to rule out the possibility that experiencers of the phenomena Jinks calls "Disappearing Object Phenomenon," (a type of personal Mandela Effect) are psychologically distinguishable from non-experiencers. That is to say that experiencers are not suffering from a variety of psychological issues and conditions more than non-experiencers. Such psychological issues include: transience, telescoping, errors of commission (suggestibility, illusionary correlation, confirmation bias), perceptual blindness (i.e. inattentional blindness), altered states of consciousness (hallucinations, autohypnosis), encoding errors ("absent mindedness"), memory distortion (pre-existing beliefs), and loose associationism. [5-7]

SIMULATION THEORY / DREAM

The Matrix movie introduced the idea that some kind of Artificial General Intelligence might be simulating our entire reality, setting the stage for us to seriously consider the idea that reality may be much more—and much weirder—than it seems. It also introduced concepts of humanity awakening, evolving, and ultimately gaining skills capable of tackling powerful forms of Artificial Intelligence (AI). As we now see ever-increasing evidence of AI, we can be understandably suspicious of how much influence these ubiquitous AI systems have in our daily lives. The weird blend of individually targeted yet impersonal advertisements predicted by Philip K. Dick has arrived, as the internet of things can now recognize us as we enter buildings.

Our increasing familiarity with a sense of relative anonymity while playing games featuring detailed artificial, simulated worlds lends additional credibility to the idea that whole new imaginary realities can be created within computer systems, which in turn makes it easier for us to imagine that perhaps we may be trapped in such a simulation.

Recent momentum for Simulation Theory comes from advances in quantum computing, and comments by people such as Geordie Rose, co-founder of D-Wave quantum computing. In 2015, Geordie asserted that parallel universes exist that can be manipulated by quantum computers, to exploit parallel universes to solve problems, stating:

> *"Science has reached the point now where we can build machines that exploit those other worlds, and quantum computers are perhaps the most exciting of all of these."* [5-8]

The simulation viewpoint was shared by Bank of America Merrill Lynch with it's clients in September 2016, stating that there is a 20%-50% chance that we're living in the matrix — and the world we experience as real is actually a simulation. This assertion has been shared by Elon Musk and Oxford philosopher Nick Bostrom, who wrote the thought-provoking paper, *"Are We Living in a Computer Simulation?"* Bostrom argued that at least one of the following proposals is true: (1) the human species will likely go extinct before attaining a "post-human" stage, (2) any post-human civilization is highly unlikely to run simulations of their evolutionary histories, and (3) we are likely living in a computer simulation. [5-9] Bostrom's high-tech overhaul of more ancient spiritual traditions of seeing the "outside world" as Maya, or illusion, seems especially attractive to people who spend time immersed in computer games and activities.

In June 2016, Elon Musk announced that the chance that we're not living in a simulation is "one in billions," citing computing technology that will become indistinguishable from real life. Yet, the qualities of our life experiences feel so much richer, more joyful, and more meaningful than anything computer simulations can offer. Computer simulations are our newest metaphor for reality, so it's natural that we would automatically think of them when trying to envision how reality might possibly be ordered and arranged.

In 2021, MIT computer scientist and computer silicon valley game entrepreneur, Rizwan Virk, published *The Simulated Multiverse,* covering the idea of how simulation theory, combined with quantum physics multiverse theory, might help to explain the Mandela Effect. Virk concludes, *"(1) That every major decision we might have made, every road not taken, was in fact taken by some version of ourselves (by a time instance). (2) That we (or someone or something) chose this particular timeline over the others for us to experience in this moment."* Virk prefers to view our roles as players in such a simulated game to be like an RPG (role player game), rather than NPC (non player characters), providing us with conscious agency within such hypothesized simulation. [5-10]

Much of the appeal of Simulation Theory arises from the idea that reality can be described purely in terms of information, which can be conveyed with amazing precision and simplicity through mathematics. Somewhere along the way from physicist Eugene Wigner's early observation in 1990 of what he called *"the unreasonable effectiveness of mathematics in the natural sciences,"* we have ended up with physicist Max Tegmark proposing in 2014 that our external physical reality is a mathematical structure. [5-11] [5-12] Tegmark asserts that our physical universe is not only described by mathematics, but it actually *is* mathematics, and he attributes the possibility of self-awareness to mathematical structures.

The most fundamental problem with simulation theory is the question of what, exactly, is being simulated, and by whom? The same people who might mock those who say our world exists on a turtle's back seem to be adopting similarly blindered logic when presenting a theory of "someone"simulating "something" "somewhere." Simulations within simulations within simulations start sounding like the kind of creation myths that skeptics have long-ridiculed as the Earth resting on a turtle, resting on a turtle, resting on a turtle—such that it's "turtles all the way down." The simulation theory is not yet complete enough to answer the question of "What is the true nature of reality?" nor does it provide us with testable hypotheses that can be confirmed or denied.

What is indisputably true about simulation theory is that our trusted senses do not fully provide us with what is actually real—instead, what we sense is entirely based on simulation. This is beautifully described by cognitive scientist and author Dr. Donald Hoffman, in *The Case Against Reality.* [5-13]

When considering the similar, yet slightly more natural theory that we exist in a Dream, some of these same questions and considerations exist, yet we can more readily envision how we might be both the Dreamer and the Dream—that we are the ones who both experience what we imagine, and also are powerful creators, whose observational focus of attention (individually and collectively) plays an essential role in selecting the realities we experience.

SCIENTISTS AT CERN

The European Organization for Nuclear Research was established in 1954 as *"Conseil Européen pour la Recherche Nucléaire"* (CERN) by 12 European governments to operate the largest particle physics laboratory in the world. The CERN laboratory is located in a northwestern suburb of Geneva, Switzerland. It is composed of 23 member states. Collaborative experiments are conducted at CERN through billions of dollars and Euros of funding flowing into CERN programs in high energy physics programs. Some wonder if strange things are afoot at the Large Hadron Collider (LHC), and whether CERN scientists opened a time portal, created a black hole, or ripped a tear in the fabric of space-time. Some find the "666" in CERN's logo and the Shiva statue at the LHC disturbing, and were disheartened by some goings-on at the LHC's opening ceremonies.

Intriguingly, CERN is now credited in some sources as being the originator of the internet. This is a Mandela Effect for some of us, who recall the internet originating from department of defense funding in the USA. In the 1980s while studying Physics at UC Berkeley, I first utilized an early inter-campus version of the internet on something called Advanced Research Projects Agency NETwork (ARPANET). This earliest version of what became the internet started in 1962 when J.C.R. Licklider became the first head of the computer research program at the Advanced Research Projects Agency (ARPA), which was a branch of the United States Defense Department. Many networks emerged in the late 1970s, with National Science Foundation Network (NSFNET) eventually replacing ARPANET as the backbone of the internet. In the 1980s, NASA and many United States Department of Energy sites had ARPANET nodes. In 1983, the term "internet" was adopted along with TCP/IP communication protocol. In 1989, Tim Berners-Lee at CERN in Switzerland proposed a hypertext system to run across the internet on a wide variety of different operating systems, ushering in the World Wide Web.

In July 2022, *Vice* magazine reporter, Jason Koebler, investigated the question of CERN's connection to the Mandela Effect. Koebler interviewed CERN spokesperson and particle physicist, Clara Nellist, who responded to concerns that CERN's Large Hadron Collider might be changing things such as the name of Double Stuffed Oreos. Nellist said, *"There are much higher energy particle collisions happening in our atmosphere all the time. What CERN is doing is tiny in comparison. I can promise you we're not going around changing the labels on your food."* [5-14]

While some express concerns that scientists at CERN may have tampered with the fabric of reality, it's worth noting that scientists working at CERN are primarily material realists, focusing on particle physics. And even if these scientists are somehow directed by dark influences, in any contest between various levels of consciousness, it makes logical sense that God's influence is always greater.

QUANTUM PHENOMENA

Since studying quantum physics at UC Berkeley, I've been impressed by the remarkable similarities between many quantum phenomena and commonly observed types of reality shifts. I've seen objects transform, transport and teleport—which may be expected in quantum computing tunneling experiments with tiny quantum particles operating at the Planck scale, but are not typically considered to be a normal part of one's everyday reality.

There is a growing awareness within the quantum physics community that as much as we may believe that "what happens in the "quantum realm" stays in the quantum realm," the very tiny realm of quantum waves and particles necessarily interacts with larger macroscopic levels of reality. Many long-standing assumptions about the nature of reality —including pillars of our modern scientific method—are being overturned, thanks to quantum physics. A majority of physicists recently surveyed agree with the idea that you and I and everyone exist in a superposition of states. Another assumption implicit to classical Newtonian physics being overturned by quantum physics involves our assumption that there exists such a thing as an objective observer, or objective truth. In 2019, quantum physics experiments proved that two trusted observational devices at the same place and same time can record two completely different events. [5-15]

Additional core attributes of quantum physics involve such concepts as non-locality and entanglement. These demonstrate that information can cross vast distances of space instantaneously. We have seen laboratory evidence showing that decisions in the future can influence events in the present and the past, and we see in quantum physics experiments how the choices of how we choose to observe absolutely influence what we subsequently witness.

The significance of quantum phenomena appearing in our everyday world is that when we take such ideas seriously, we acknowledge that we can expect to occasionally witness such things as large-scale objects appearing, disappearing, transporting, and transforming—as well as changes in the way we experience time. In fact, over more than twenty five years, exactly these things have been observed in thousands of first-hand reports documented at www.realityshifters.com.

QUANTUM IMMORTALITY

Another theory to explain what is going on with the Mandela Effect involves variations of the idea that none of us can actually die, and perhaps we might already be dead. This notion originated as a thought experiment originally called "quantum suicide," first published by Hans Moravec in 1987, and independently developed by Bruno Marchal in 1988. Physicist Max Tegmark explored the idea further in 1998, attempting to distinguish between the Copenhagen interpretation of quantum mechanics and the Hugh Everett Many Worlds Interpretation via a variation of the Schrodinger's cat experiment, from the cat's point of view. The theory of "quantum immortality" refers to the subjective experience of surviving quantum suicide, regardless of the odds. [5-16]

The quantum immortality theory took on new legs with people observing the Mandela Effect, with people wondering if we might have died in some alternate reality. YouTuber SMQ AI presented the idea that one possible explanation for the Mandela Effect and misalignment between our memories and historical recorded facts might be that a mass extinction event occurred in 2012 that killed us all, and our experience of reality is actually the afterlife. SMQ AI states, *"We died in a mass extinction event in 2012. Consensus will continue to dissolve as the Mandela Effect intensifies. We are all Cymatic echoes or vibrations that are in the afterthought of our extinction."* [5-17]

While quantum immortality theory resonates with some Mandela Effect experiencers, it leaves many unanswered questions. If we all died in 2012 and are echoes of our former selves, then why are babies still being born, and why are we learning new things, meeting new people, and starting new creative projects? Clearly any sense of fresh beginnings should not be possible, if we are already dead. And while

121

some who experience death, or who have near-death experiences (NDEs), may also experience the Mandela Effect, there are many experiencers who have not died or nearly died.

TIME TRAVELERS

The idea that time travelers were involved with the Mandela Effect was radio broadcaster Art Bell's favorite theory. Some Mandela Effect experiencers look past the bias of a forward "arrow of time," suspecting that time travelers may have somehow tampered with reality. Those subscribing to this theory point out that the small, mostly insignificant kinds of changes observed are exactly what we would expect to see if someone was testing out time travel technology. Some theorists posit that secret government agencies may be conducting time travel experiments and research, resulting in occasional glitches.

If time travelers have been modifying history, what might their motives be? Could they be mining time loops, seeking desired resources? If such things are going on, would there be some kind of irrefutable, undeniable evidence such had occurred? Actually, it seems such meddling might occur with minimal, if any indications at all. Physicist Yakir Aharonov points out an important aspect to keep in mind with regard to the future influencing the past:

> *"The future can only affect the present if there is room to write its influence off as a mistake." – Yakir Aharonov*

Time travelers might thus inadvertently or intentionally create discernible Mandela Effects through stealthy, ninja-like interventions, but there would likely always be some room for doubt—some plausible deniability built in. Closed time-like curves can provide for retrocausal action, in such a way that avoid self-contradictory grandfather paradox situations, so time travelers don't accidentally kill their ancestors and cancel out their own existence. A closed time-like curve (CTC) is a path of spacetime that returns to its starting point, predicted in 1937 by Willem Jacob van Stockum [5-18] and further elaborated upon by mathematician Kurt Godel in 1949. [5-19] David Deutsch theorized in 1991 that paradoxes created by CTCs could be avoided at the quantum scale—since quantum fundamental particles follow only fuzzy orders of probability, rather than strict determinism. [5-20]

Allowing for the inclusion of future influence on the past thus begins to seem possible, with quantum systems providing their requisite environs of plausible deniability. Put more simply, we can expect that we will likely see some retrocausal, time-reversed effects, as long as no absolutely definitive evidence exists.

Fred Alan Wolf's book, *The Yoga of Time Travel,* describes how yoga adepts can overcome five barriers to reaching a state of ego-less mind, in order to travel through time without any time travel device—no time machine is needed. Wolf points out that time travel is required by our current quantum models of the physical world, to the point that the burden of proof now resides in proving any kind of physical law exists that forbids time travel. [5-21]

People who've experienced reality shifts are much more open to considering the possibility that the future and present can change the past, lending credence to physicist John Cramer's Transactional Interpretation of quantum physics. Considering that space-time is a continuum, and physics fully allows for time-forward and time-reversed causality, a whole new world of possibility opens up with the notion of a "handshake" between future and the past allowing for the kind of symmetry in time we expect in space.

Another key point regarding time comes to us courtesy physicists Stephen Hawking and Thomas Hertog:

"Quantum physics forbids a single history." [5-22]

The enormity of impact from this elegantly modest statement suggests that we never have only one agreed-upon history. We tend to not be aware of multiple histories, nor do we often accept the possibility of their existence. Yet, when we think about how it would feel to be influenced by the future, we realize that we could expect to experience such things as: Déjà Vu, future memory, precognition, premonitions, intuitive hunches, synchronicity, and exceptional situational awareness, as well as spontaneous remissions from injury and disease.

When viewed in combination with evidence suggesting that quantum delayed choice is quite real, it's clear that we may well be the time travelers we have been looking for.

CONVERGING TIMELINES / REALITY BUBBLES

There is a theory that at some point in time, such as in the year 2012, realities converged. I saw visions of this being the case, which I wrote about in the anthology, *2012: Creating Your Own Shift.* [5-23] This idea of converging realities can be clarified in a thought experiment, since we are not literally crashing into another world or worlds. If we were, we might expect to see something along the lines of what transpired on

123

an episode of the TV series, *Fringe*, where parallel possible worlds literally collided, to catastrophic physical effect.

There is a fine art to observing adjacent and converging timelines. Our problem-solving ability relies upon our awareness of distinctions between what "actually" happened and what we "expect," so by observing more about both of these areas, we can glean insights about Mandela Effects. For example, we might see groups of people with similar "Mandela Effects" grouped geographically who remember Nelson Mandela's death—but other times, we might observe that groupings of observers are based on something other than geography, such as area of professional expertise.

The concept of time "lines" may not be best for contemplating what is going on with the Mandela Effect, since few people completely match memories with others, and some shifts "flip-flop" back and forth. With regard to various reality shifts and Mandela Effects—it's unlikely that any two people would find all their memories matching anyone else's. We instead witness what seems more like subjective realities. The notion of "consciousness soup" might convey the remarkable ways that our memories and realities appear to be subjectively our own amidst a shared collective, as Shane Robinson discussed with Candace Craw-Goldman. [5-24] Amidst this larger collective consciousness exist "reality bubbles" or smaller cones of experience shared by individuals.

Author Roger Marsh presents a complementary view of the idea of reality bubbles in his book, *TruthBubble*. Marsh suggests that we exist in subjective realities, and encourages us to rise above petty fear-based limitations, toward harmonious, balanced evolution of consciousness amidst uncertainty. [5-25]

NONLINEAR ORTHOGONAL TIME SHIFTS

Credit for the idea of nonlinear time rightly belongs to indigenous peoples. Glenn Aparicio Parry's book, *Original Thinking*, explores the concept of living with an indigenous, nonlinear relationship to time and space, where even the meaning of the word "original" indicates all that exists through all time. Time is not linear to indigenous peoples, and locality and spatial separation is recognized as an illusion. Within such a nonlocal, interconnected framework for reality, we can appreciate how connections between everyone and every thing can be felt and known. [5-26] There is shared awareness in first nations communities in accordance with quantum physics, that came to light in discussions with

American physicist David Bohm, Blackfoot Leroy Little Bear, American linguist Dan "Moonhawk" Alford, and many others in Kalamazoo, Michigan in a first meeting hosted by the Fetzer Institute in 1992. Some shared points of agreement that were recognized as first principles included: (1) everything that exists vibrates, (2) everything is in flux (the only constant is change), (3) the part enfolds the whole (not just whole is more than sum of its parts), (4) there is an implicate order to the universe, (5) the cosmos is basically friendly, (6) Nature can be taught new tricks, (7) what physicists call quantum potential corresponds to what Native Americans call spirit, and (8) the principle of complementarity. [5-27]

Dan Moonhawk Alford described himself as hanging out *"at the lonely intersection of language, physics, Native America and consciousness,"* The precious hours I spent with him at my first Language of Spirit conference in New Mexico were life-changing. Moonhawk had a gift for illuminating the importance of the ways language shapes consciousness, and a genius for inspiring people to think differently. He pointed out how in Nature, *"A does not always equal A,"* because *"you can never step into the same river twice."* Modern science and logic was constructed with a mechanistic bias, while the natural world is ever-changing, generous, alive and profound. Moonhawk emphasized that our English language is noun-based, affecting our view of the cosmos. Some Native Americans can talk for hours or even an entire day without once uttering a single noun. Such a concept is mind-blowing to most westerners who are accustomed to focusing on things with the biased assumption that nouns exist everywhere. The idea of everything in reality being active, flowing, moving, and changing is radically different from the typical modern reductionist viewpoint, yet is a perfect match for quantum physics.

How might the idea of nonlinear orthogonal time lend itself to experiencing the Mandela Effect? American author Philip K. Dick (PKD) wrote science fiction books and stories including *Man in the High Castle*, which envisioned the world if Germany had won World War II. In his 1977 speech in Metz, France, entitled, "If you find this world bad, you should see some of the others," he shared how we might perceive changes taking place outside of linear time, positing something he called "orthogonal or lateral change." PKD said:

> *"Contemplating this possibility of the lateral arrangement of worlds—a plurality of overlapping Earths along those linking axes a person can somehow move—can travel in a mysterious way from worst to fair to good to excellent. Contemplating this in theological terms, perhaps we could say that herewith we*

suddenly decipher the elliptical utterances which Christ expressed, regarding the Kingdom of God. Specifically, where it is located." [5-28]

Jesus is reported to have said, *"My kingdom is not of this world,"* adding, *"The kingdom is within you,"* or, *"it is among you."* PKD invites us to consider that Christ was teaching his disciples the secret of crossing this orthogonal path to literally reaching the most wondrous of possible worlds. PKD continues:

> *"Christ and St. Paul both seem to say emphatically that an actual breaking through into time, into our world, by the hosts of God, will unexpectedly occur. Thereupon, after some exciting drama, a thousand year paradise, a rightful kingdom, will be established—at least for those who have done their homework and chores and generally paid attention... have not gone to sleep, as one parable puts it."*

PKD describes his books as being based on *"fragmentary residual memories of such a horrid slave state"* in the USA. He remembered three parallel, alternate lives. One life was deeply traumatizing for him, inspiring his works of fiction about the dangers of dictatorial regimes, and remembering a *"black iron prison world that had been."* PKD experienced a second world as *"our intermediate world in which oppression and war exist, but have to a great degree been cast down."* A third world was filled with beauty, peace, and love as *"a third alternate world that someday, when the correct variables in our past have been reprogrammed, will materialize as a superposition onto this one, and within which, as we awaken to it, we shall suppose we always lived there, the memory of this intermediate one, like that of the black iron prison world, eradicated mercifully from our memories."* In 1977, he said he knew of no one who had publicly claimed experiencing simultaneous, alternate, parallel lives. While PKD may not have met them, such people surely did exist. Since I began researching this phenomenon in 1998, I received numerous reports from people experiencing living very different present lives—some observing parallel realities back in the 1970s and earlier.

The idea within physics of closed time-like curves (CTC) has been conceived as providing the means by which one might travel backwards in time, or to other adjacent possible universes. [5-29] Physicist Fred Alan Wolf points out that classical "mixed states" arise when an "observation" is said to occur in closed time-like curves, resulting in a so-called reduction of the quantum wave function, and the appearance of a classical world. Wolf says:

"In brief, Deutsch's CTC nonlinear post quantum physics model may represent the action of a conscious mind." [5-30]

SPIRITUAL APOCALYPSE

The meaning of the word "apocalypse" comes from a Greek word meaning revelation, such as in an unveiling of previously concealed truth. Some views of apocalyptic times envision an end of life as we know it—a literal "end times." Some Mandela Effect experiencers gravitate toward this explanation as to what is happening, citing prophecies found in spiritual texts, such as Daniel 7:25 in the Bible, stating that the Antichrist will *"change times/seasons and laws."*

When we consider that the Kingdom of God might be within us, we can imagine how we experience varying degrees of goodness in realities. Philip K. Dick (PKD) proposes that Christ may have had in mind what he calls *"the lateral axis of overlapping realms, which contain among them a spectrum of aspects ranging from the unspeakably malignant to the beautiful."* Both Jesus Christ and Saint Paul emphatically state that a breaking through into time into our world by the hosts of God will *unexpectedly* occur. A rightful kingdom will unexpectedly be established for those who have not "gone to sleep," as one parable puts it, because for the rightful and faithful, it is always daytime; the Kingdom will not be visible to those outside it. PKD explains that:

> *"Some of us will travel laterally to that better world, and some will not. They will remain stuck along the lateral axis, which means that for them the kingdom did not come—not in their alternative world—and yet meantime it did come in ours."*

PKD recognized the potential existence of overlapping alternate worlds, such that,

> *"events on one track cannot be perceived by persons not in that track."* [5-31]

Intriguingly, PKD noted that,

> *"though originally I presumed that the differences between these worlds was caused entirely by the subjectivity of the various human viewpoints, it did not take me long to open the question as to whether it might not be more than that. That in fact, plural realities did exist superimposed onto one another like so many film transparencies."*

With remarkable prescience, PKD doubted whether we would:

> *"ever be able in any real fashion to demonstrate—to scientifically prove—that such lateral change processes do occur. Probably all we would have to go on*

would be vestiges of memory—fleeting impressions, dreams, nebulous intuitions —that somehow things had been different in some way, and not long ago, but now."

PKD mentions examples of reality shifts similar to what I've experienced and documented, such as reflexively reaching in the wrong place for something like a light switch or air vent—or turning a timer dial the wrong direction, as I describe in my book, *Reality Shifts*. [5-32]

SCHUMANN RESONANCE

Correlations between the Mandela Effect and spikes in the Schumann resonance have been noted by Mandela Effect experiencers, who notice that increases in reality shifts seem connected with these surges. Before 2014, the Schumann Resonance rarely rose above 15-25 Hz, and after 2014, it regularly reached frequencies above 36.

In 1951, German physicist Professor Winfried Otto Schumann predicted that there exist electromagnetic standing waves in the atmosphere, within the cavity formed by the surface of the Earth and the ionosphere. Schumann and his students at the Technical University of Munich calculated that this frequency ought to be about 10 Hz. This so-called "Schumann Resonance" was published in *Technische Physik* where it was seen by Dr. Ankermueller, a colleague of Hans Berger, inventor of the EEG machine, who measured human brain waves at 7.8 to 13 Hz. [5-33]

In 1954, measurements by Schumann and König detected resonances at a main frequency of 7.83 Hz, which seemed fairly stable. The spherical earth-ionosphere cavity between the conductive surface of the earth and the outer boundary of the ionosphere is separated by non-conducting air. When electromagnetic impulses generated by electrical discharges (such as lightning) spread laterally into the cavity, the discharges have a "high-frequency component" between 1 kHz and 30 kHz, followed by a "low-frequency component" consisting of waves and frequencies below 2 kHz. This produces electromagnetic waves in very low frequency (VLF) and extremely low frequency (ELF) ranges.

In April 2020, Jerry Hicks observed a correlation between a global meditation conducted on April 4, 2020, a discernible spike in the Schumann resonance, and a fresh batch of newly observed Mandela Effects. Some of the meditations occurring in this time period include multitudes of the 1.3 billion Indian citizens asked to pray by Indian Prime Minister Narendra Modi. Many other meditation groups joined in on this worldwide mass meditation event, and a brilliant spike appeared when these meditations were conducted. [5-34]

GREAT AWAKENING:
EVOLUTION OF CONSCIOUSNESS

Many within the Mandela Effect community feel we are experiencing changes in human consciousness on a collective and individual scale. After researching the Mandela Effect and reality shifts for over 25 years, it's clear to me that false memories cannot account for all of the shifts that I and others have seen. The scope and scale of the shifts are far larger than any conspiracy, they affect huge numbers of people, and many people are remembering very similar things.

Indigenous people have words to describe how consciousness can change the world, such as the Hopi word, *tunatyava*, meaning "comes true, being hoped for." Hopi legends describe how the world has ended several times, most recently with a global flood, and provides tips by which we can find our way through transitional times, accompanied by lights that can be followed by night, and "clouds" that can be followed by day—and providing us with advice for how to successfully move to the next higher level of consciousness during such times of global change by keeping open hearts and open crown chakras. [5-35]

Some observe that the end of the Mayan calendar corresponds with a pivotal moment of increases in the Mandela Effect. Swedish physicist Dr. Carl Calleman describes *The Nine Waves of Creation*—waves of consciousness corresponding to key epochal moments in world and human history. With the appearance of each wave, previous waves remain and combine with the arrival of new information and technologies. At the dawning of the Eighth Wave, humanity received the invention of the cell phone. The Mandela Effect seems to qualify as anticipated new technology arriving with the advent of the Ninth Wave of high frequency consciousness, starting in October 2011. [5-36]

A surge of interest in the Mandela Effect after 2010 was preceded by increased popular interest in meditation, which may have had an effect on global consciousness. We reached a tipping point somewhere between 2012 and 2015, when surveys indicated that for the first time in history, 10% of Americans meditated regularly. The American Psychological Association (APA) survey indicated that we reached 10% in 2012 [5-37]; the American CDC's National Center for Health Statistics reported that the percentage of Americans regularly meditating rose from 4% in 2012 to 14% in 2017. [5-38]

In 2014, Google's most popular employee course (with a six month waiting list) was "Search Inside Yourself." Participants realized that putting ideas from this course into how they live their lives improved the quality of their relationships, noting, *"I take time to think through things and empathize with other people's situations before jumping to conclusions."* [5-39]

Time magazine featured meditation in a 2014 cover story, "The Mindful Revolution: The Science of Finding Focus in a Stressed-Out, Multi-Tasking Culture." Average Americans are increasingly meditating to reduce stress, relieve pain, improve relationships, and experience a greater overall sense of well-being in their lives. As mindfulness goes mainstream, people of all ages and walks of life make meditation a regular daily part of their lives. [5-40] Meditation is acknowledged as a winning sports strategy, too. The Seattle Seahawks won the 2014 Super Bowl with offensive tackle Russell Okung quoted by ESPN as saying,

> *"Meditation is as important as lifting weights and being out there on the field for practice."* [5-41]

One of the first people to describe higher levels of conscious awareness in mathematical terms was Gottfried Wilhelm Leibniz, one of the inventors of Calculus. He viewed consciousness as requiring both a first order perception of some "x" and a second order reflective perception of the original perception of "x." Leibniz says:

> *"It is good to make a distinction between 'perception,' which is the internal state of the monad representing external things, and 'apperception,' which is 'consciousness' [conscience], or the reflective cognition of this internal state, which is not given to all souls, or at all times to the same soul."* [5-42]

Leibniz shared an additional insight to the nature of how conscious perception arises—that it does so gradually, *"by degrees from [perceptions] that are too minute to be noticed"* —i.e. Unconscious perceptions. [5-43]

In November 2022, I talked with synchronicity expert, Dr. Bernard Beitman, about meaningful coincidences, adventures in the psychosphere, and memorable experiences with dogs. Beitman acknowledged that synchronicity shaped his life, and led him to explore how it operates. Beitman discovered that the psychosphere provides a kind of mental atmosphere where a person's need defines and clarifies an intention, so the mind can arrive at the needed information in extraordinary, wonderful ways. *"Coincidences are not commands, they are suggestions,"* Beitman said, since we can identify underlying patterns in life that we recognize as synchronicity, serendipity, or simulpathity–such that we might, for example, experience the feelings of a loved one who is not with us at the time. [5-44]

The heart of Beitman's concept of "psychosphere" is the idea of a mental atmosphere surrounding us in much the same way as the air we breathe, in which all beings are immersed. Beitman credits earlier authors for having contributed to the idea of a psychosphere, including the psychiatrist, Ian Stevenson, and also French philosopher and Jesuit priest Teilhard de Chardin and the bio-geochemist Vladimir Bernatsky, who talked about the *noosphere*, meaning a sphere of mind.

Dr. Beitman's book, *Meaningful Coincidences,* provides inspiration for exploring the psychosphere, as well as first-hand accounts of people experiencing synchronicity, and four personality traits share by "Coinciders," who often experience such phenomena in their lives:

(1) Self-referential thinking, characterized by beliefs that "events around me refer and have to do with me;"

(2) High emotional charge (strong positive or strong negative);

(3) Commitment to the idea that God intervenes personally in our lives; and

(4) Searching for meaning. [5-45]

Real synchronicity helps us feel that something special is going on, when we note events happening together, at the same time. What is it about synchronicity that so touches our imagination, hearts, and souls? This universal experience has the power to spark a sense of reverence and deep curiosity. While those who experience synchronicity can know how profoundly moving and life-changing it can be, Beitman's examples of meaningful coincidence inspire us to feel a sense of awe and wonder that there may be something much greater than ourselves. Swiss Psychologist Carl Gustav Jung coined the term *synchronicity* to signify "*the simultaneous occurrence of two meaningful but not causally connected events,*" or as "*a coincidence in time of two or more causally unrelated events which have the same or similar meaning... equal in rank to a causality as a principle of explanation.*" Synchronicity seemed to Jung to introduce evidence of a higher order of consciousness operating in our lives—such as the time when a scarab beetle, unusual for that area, came to tap on the window exactly when Jung and a client were discussing how the young woman had dreamt of being given a golden scarab.

Synchronicity is the occurrence of a physical event in the world which occurs at or near the same time that it is being discussed or thought about. The essence of synchronicity is felt, for there is often personal significance and meaning associated with it. When you consider that this kind of interplay is one way that your identity as eternal spirit and your physical self engage in dialogue, you may find yourself looking forward to experiencing more frequent synchronicity in your life. Beitman emphasizes that the fundamental lesson of meaningful coincidences is that each human mind is much more connected to other minds and its environment than current scientific thinking suggests. By tracking coincidences, we can construct maps of and go adventuring in the psychosphere.

While some people may consider the notion of instantaneous interpersonal connection to seem far-fetched, Beitman shares examples of how people have gleaned insights as to how simultaneous discoveries have been made of remarkable breakthroughs by different people in different places on the planet at roughly the same time—such as the idea of biological evolution. In 1922, William Ogburn and Dorothy Thomas documented 148 major scientific discoveries that were simultaneously made by two or more people around the same time. [5-46]

Beitman provides both his own viewpoint of how some synchronous events occur, as well as the perspective of those who were experiencing them. For example, one woman described how she had found information that she needed at the right place and time to save her father's life. Beitman shared his belief that this woman had subconsciously registered her father's distress at a distance, and then as she was reading a book was able to know that she had found information that fit her father's situation—or she may also have been aided by an intelligence outside of herself in the psychosphere.

Psychologist Igor Grossmann has conducted studies that suggest we can improve the clarity of our analytical reasoning and increase wisdom by practicing the ancient art of *illeism*. [5-47] A person practicing *illeism* receives a kind of self-identity "level up" by speaking and writing about themselves consistently in the third person. For example, a journal entry about my day might read something like,

> *"This morning, Cynthia was grateful to have been able to take a walk in nature before getting started with her responsibilities for the day."*

By identifying as the one observing the human being who is thinking, feeling, speaking and acting at a higher level of conscious agency, we gain benefits of more fully embodying our eternal, infinite, wise higher selves, rather than over-reacting to events unfolding at any given moment. We can thus find it easier to rise above temptations to fall back into old habitual emotional patterns, and embody the qualities we wish to see in the world.

Through recognition of the Mandela Effect, humanity is accepting an invitation to think outside the Boolean true-false box, and enter into a wonderfully interconnected, dynamic reality based in quantum consciousness, demonstrated and evidenced on a macroscopic scale. We now witness large-scale evidence of such quantum phenomena as: non-local effects, superposition of states, entanglement, teleportation, bi-causality (going forward and backward in time), and tunneling.

Evolution of consciousness is suggested by the way we witness such gentle Mandela Effects, such as changes in foods. Seemingly minor changes such as Haas avocados now always having been known as "Hass," (despite still being pronounced the other way) and Brussels Sprouts becoming more palatable (as the popular online comic strip XKCD pointed out in 2019) seem largely benevolent. [5-48]

Evolution of consciousness is implied in the way that many Mandela Effect changes can be recognized as evolutionary improvements, such as the kidneys having moved up from the old "kidney punch" danger area, into the relative safety of the lower rib cage. In another improvement, our hearts are now situated more in the center of our chests, so we can accurately have our blood pressure taken on either arm. Sociopaths now have empathy, and treatment for Alzheimer's now exists. We can increasingly find solutions to most any problems we know we most need answers to, regardless of severity and scale.

REVIEWING MANDELA EFFECT THEORIES

An intriguing aspect about all Mandela Effect theories is that when we pursue them far enough, we end up at pretty much the same place: ultimately, consciousness is somehow influencing the physical world. My personal favorite Mandela Effect theory is a hybrid combination of awareness that reality is like a dream (or simulation) with quantum phenomena occurring at every scale, and with evolution of human consciousness involving levels of conscious awareness and activation of higher levels of conscious agency. Like young children exploring the wonders of the natural world with kindness and cooperation, it's possible that the Mandela Effect offers us the opportunity to interdimensionally walk, skip, and jump through parallel alternate worlds in space and time.

* * * * * * *

The next chapter, *Science of the Mandela Effect,* delves into some of the science behind the Mandela Effect, and includes a glossary of Mandela Effect scientific terms, and a table of ideas that are similar between Mandela Effect scientific terms and Mandela Effect terminology.

~~~~~~~~~~~~~~~~~~~~~~~~~~~~~~~~~~~~~~~

EXERCISE:
CHOOSE YOUR FAVORITE THEORY

What do you think might be causing the Mandela Effect? Write down what feels right to you. Feel free to combine two or more theories, and elaborate further to add meaningful aspects to whichever theories feel important and meaningful to you. What about your favorite theory feels most resonant to you? What does it explain better than other theories?

~~~~~~~~~~~~~~~~~~~~~~~~~~~~~~~~~~~~~~~

EXERCISE:
START AN ILLEISM JOURNAL

Practice the ancient art of *illeism* and obtain a self-identity "level up" by thinking and writing about yourself in the third person. Think about three key events that happened over the past day and describe them from a higher level of yourself. The goal of this exercise is to practice the art of stepping back to observe your thoughts and feelings—identifying less with your daily egoic self, battered by life's drama, and more with the one observing life from a higher vantage point. Practice writing something like: *"Once she awakened and dressed, Cynthia tended to plants in her garden, feeling refreshed by the morning sunshine and bird song."*

~~~~~~~~~~~~~~~~~~~~~~~~~~~~~~~~~~~~~~~

EXERCISE:
THINK IN VERBS, WITHOUT NOUNS

Just for fun, see if you can write a meaningful sentence without any nouns, with as many verbs as you like. Imagine that this is how Nature truly works—constantly changing and flowing, connecting and relating. Envision yourself as pure beingness, no longer constrained to feeling like a "thing," but rather engaged in doing whatever activities you enjoy. Observe your beingness free from any expectations, judgments, or criticisms, with gentle awareness that rest and daydreaming are valuable activities, too.

~~~~~~~~~~~~~~~~~~~~~~~~~~~~~~~~~~~~~~~

*"The wave nature of an indefinite state
captures the psychological experience
of conflict, ambiguity, confusion, and uncertainty;
the particle nature of a definite state
captures the psychological experience
of conflict resolution, decision, and certainty."*

—*Jerome Busemeyer & Peter Bruza*

Chapter 6

SCIENCE OF THE MANDELA EFFECT

We're in the midst of a quantum invasion that promises to influence every aspect of our lives. As quantum physics boldly infiltrates every branch of science, outdated worldviews and assumptions are tossed aside in its wake. We now know that quantum processes are proven to occur in places where many scientists never expected to find them, such as in warm, wet, noisy biological systems, where they play pivotal roles. We are privileged at the dawning of this new Quantum Age to see how quantum logic and quantum processes provide optimal solutions while transforming our view of ourselves and the cosmos. Physicists developing a Theory of Everything are working with a fundamental assumption that quantum mechanics operates everywhere and at every scale. Quantum phenomena occurs in such a way that observers can't be separated from what is being observed. Consciousness plays an essential role, such that true objectivity from being "on the outside" or "above it all" is essentially impossible. We are both observers and experiencers of Mandela Effects, and those of us asking questions about these immersive experiences often notice amazingly nuanced, personal shifts. Nature answers questions that we feel most strongly about in this participatory cosmos, providing us with unique, specific experiences based on questions we ask.

In the science of quantum physics, we find that seemingly strange quantum qualities correspond to our otherwise seemingly illogical human logic. As Jerome Busemeyer and Peter Bruza state in their book, *Quantum Models of Cognition and Decision*, *"The wave nature of an indefinite state captures the psychological experience of conflict, ambiguity, confusion, and uncertainty; the particle nature of a definite state captures the psychological experience of conflict resolution, decision and certainty."* We can thus begin to grasp some of the advantages intrinsic to our fuzzy, intuitive thinking, and how our analytical thinking can enhance clarity. [6-1]. In addition to helping us understand how we think, quantum physics helps to explain how life as we know it can even exist at all, since it is necessary for basic biological functions. [6-2]

The idea that the many worlds of quantum physics might be one and the same as the multiverse has been proposed by physicists Yasunori

Nomura and Raphael Bousso of UC Berkeley, as scientists anticipate finding evidence that we are living in a multiverse. By viewing the many worlds of quantum physics as one and the same as the multiverse, we can envision jumping to a parallel possible reality via Einstein-Rosen wormhole bridges, existing as miniature black holes. An artist's rendition of this appears on the cover of my book, *Quantum Jumps*.

Most types of reality shifts have quantum physics counterparts. For example, Flip-Flops, where something has first "always been" one way, and then suddenly "always been" another is similar to witnessing superposition of quantum states. Mandela Effects can thus be viewed as completely natural processes.

SEEING TRUE REALITY

What if the world we see isn't fundamental reality? Our brains filter everything we sense, presenting simple, incomplete interpretations. Donald Hoffman is a professor of cognitive science, computer science, and philosophy at UC Irvine, who likens the way our brains process information to icons on a computer desktop. We can click and move an icon representing a file to an icon representing the Trash, and expect that file will be deleted—yet what the computer does beyond the user interface is quite different and more involved. Similarly, our brains translate raw sensory perception information into interpretations of sight, smell, taste, touch, and sound. Hoffman explains, *"… to experience is to construct, in each modality and without exception,"*

Quantum physics and quantum biology studies confirm this, with ever-increasing scientific evidence showing this is true. [6-3] [6-4] And the implications of this reconstruction are profound. Not only do our senses provide an incomplete view of reality, they can also mislead us into assuming that only what can be measured is real.

Donald Hoffman recognizes the possibility that reality might be better described beyond the confines of our current spacetime construct, perhaps in mathematics based on the *Amplituhedron,* which is a geometric structure first proposed by Nima Arkani-Hamed and Jaroslav Trnka, that enables simplified calculations of particle interactions in some quantum field theories. Rather than being fundamental, spacetime might merely be our mind's way of making sense of infinite inter-dimensional reality.

This current dawning of the Quantum Age is an ideal time to reconsider old scientific assumptions of material realism, and to replace them with quantum ideas of: subjectivity, entanglement, superposition, conscious observation, tunneling, teleportation, and "spooky" action at a distance. Thanks to quantum physics, we see evidence that reality is not entirely comprised of matter, that wholes cannot always be reduced to smaller parts without losing harmonious integrity, and that actions can occur non-locally. Not everything can be accurately measured—in many cases we can only estimate with probabilities. Different observers at the same place and time can witness different events. Effects are observed to precede causes. Taken together, these are revolutionary changes to our fundamental worldview. Thanks to seeking understanding of the Mandela Effect, we now have practical, everyday reasons to embrace new quantum thought.

QUANTUM PROCESSES ARE NATURE'S WAY

Nature herself appears to be fundamentally quantum, with certain aspects of quantum physics never admitting a classical understanding. The role and domain of classical logic seems to be intrinsically limited —as it does not yet incorporate such truths as subjective perspective literally changing the physical world, as experimental studies demonstrated in 2019, where observational devices can witness different events viewed at the same place and time. [6-5][6-6]

Some of the world's leading experts in the field of quantum computing, such as Scott Aaronson, sound enthusiastic regarding the imminent arrival of "quantum supremacy"—the expectation that quantum computing at some point will out-perform classical computing by a significant margin under verifiable, reproducible conditions.

When I think of the phrase "quantum supremacy," the first thing that comes to my mind is the title of a wonderful paper published in 2015 by physicists David Jennings and Matthew Leifer, *No Return to Classical Reality.* This paper begins with the audacious words,

"At a fundamental level, the classical picture of the world is dead, and has been dead now for almost a century."

Jennings and Leifer continue,

"We now have a range of precise statements showing that whatever the ultimate laws of Nature are, they cannot be classical." [6-7]

If we're no longer in Kansas (i.e. classical reality), we can expect to occasionally witness such quantum phenomenon as "spooky action at a distance," and superposition of states.

PRIMACY OF QUANTUM LOGIC IN THE NATURAL WORLD

"Nothing is black or white."
—*Nelson Mandela*

The typical materialist view of analytical logic is only partially complete and partially true. Quantum logic shares similarities with Asian four-fold logic, which includes binary categories of *True* and *False*, as well as even larger categories of *True-and-False* and *Not-True-Not-False*.

Quantum computers depend on quantum logic, just as classical computers depend on True/False Boolean logic. Boolean logic was officially named in 1847 in conjunction with Charles Babbage developing the first programmable device with his Babbage computer—which was the ancestor of modern computers in our daily lives. Early computers depended upon George Boole's Boolean logic beginning in the 1800s for obtaining consistent answers.

Quantum logic will necessarily take precedence for future computing systems, as a whole new conception of what is logical and rational becomes globally appreciated. Many assumptions that seem logical and rational according to True/False Boolean logic are far too limited for quantum logic—including such assumptions as: we only need to pay attention to material things, nonlocal effects can be ignored, there is such a thing as an objective and uninvolved observer, and we can have precise answers to problems. Quantum logic overturns all these assumptions while adding possibilities of being true and false, not-true and not-false, and also adding the exciting development of two-way logic gates through time.

How can we best appreciate quantum logic? Occam's Razor invites us to avoid overly complex explanations. William of Ockham's words on this matter were, *"Pluralitas non est ponenda sine necessitate,"* which translates to "entities should not be multiplied unnecessarily." Curiously, and appropriately for our topic, much controversy occurs when scholars attempt to provide proper attribution for this quote, as noted in an article called, *The Myth of Ockham's Razor*, at the Logic Museum. It appears there might currently be no physical evidence of any kind to support that William of Ockham is the true originator of the idea behind Occam's Razor, so Occam's Razor itself appears to be an example of the Mandela Effect. [6-8]

Quantum logic has much to do with synchronicity. Synchronicity is the occurrence of a physical event in the world near the same time it is being discussed or thought about, and the essence of synchronicity is felt, as there is often significance and meaning associated with it. While synchronicity might seem illogical according to classical Boolean logic, it appears completely logical when viewed from a quantum logic perspective. Quantum logic may reveal an interplay through levels of consciousness to us through coincidences in "internal" thought and "external" action as we play with our sense of identity based on how we relate to the cosmos via levels of conscious agency, as Leibniz suggests.

Another surprising aspect of quantum logic is that it demonstrates that we cannot ever fully know all facts, since some information exists in a kind of blurry possibility field, reminiscent of the notion of plausible deniability. In plausible deniability, an individual can truthfully deny knowledge of any particular truth that may exist because the subject is unaware of that information. When we combine uncertainty with the way that quantum logic gates operate backwards and forwards through time, we begin to see a quantum mechanism at work by which future observations and choices influence the past. Facts are not entirely fixed, and only awareness—consciousness—remains eternal.

Nature, like consciousness itself, has a wild, unpredictable, nonlinear quality to it. This playful, untamed reality functions with occasionally changing "facts," such as whether or not American Thanksgiving has always fallen on the third or fourth Thursday of November, whether Nelson Mandela died in prison or after having been president of South Africa, whether the Berenstain Bears were once known as Berenstein, or whether William of Ockham really was the first to popularize the notion of *"Pluralitas non est ponenda sine necessitate"* in the Middle Ages. In Quantum Logic, the past and future can both change.

Support can be found for the primacy of quantum logic in the cognitive sciences, where researchers recognize quantum logic in such areas as: the subconscious, decisions involving unknown interconnected variables, memory, and question sequencing. The field of quantum cognition includes a set of principles from quantum theory that can clarify otherwise baffling behavioral phenomena in human decision-making. Quantum theory provides possible explanations for so-called "irrational" decision making, conjunction and disjunction probability judgment errors, over and under-extension errors in conceptual combinations, ambiguous concepts, order effects on probabilistic interference, interference of categorization on decision making, attitude question order effects, and more. Choices that people actually make in the Prisoner's Dilemma and two-stage gambling decisions are much better explained by quantum probability than classical, and such things as nonverbal cognition and memory are more clearly delineated with quantum theory and logic as well. [6-9]

Thanks to having researched reality shifts and Mandela Effects since 1998, I view classical reality as being a special case and subset of a greater quantum reality. I'm delighted that we're seeing ever-increasing evidence from scientific researchers in support of Jennings and Leifer's bold assertion that *"whatever the laws of Nature are, they cannot be classical."* With Nature's clear affinity for quantum functionality, it appears that humanity is receiving a gift of awareness of the Mandela Effect in this new Quantum Age.

WE ARE QUANTUM BEINGS

Evolutionary science is full of missing links and discontinuities where scientists had expected to see evolution from one species to the next proceed in orderly curves. Natural evolution instead typically proceeds along the lines of punctuated equilibrium, more like quantum stair steps than a smooth, classical ramp. In addition to the existence of numerous missing links in evolutionary biology, there is increasing experimental evidence of instantaneous, spontaneous evolution.

The assumption that organisms primarily evolve via random mutations was seriously shaken in 1988 by Harvard professor John Forster Cairns's suggestion that the organisms themselves preferentially produce beneficial mutations. Cairns experiments with E. coli bacteria demonstrated that in times of stress, when the bacteria were starved of food they were capable of digesting, they made *adaptive mutations* to receive nutritive sustenance from a food source they'd never been able to consume before. This phenomenon in which E. coli mutated exactly the most optimal genes precisely when that mutation was required had no known basis nor explanation from established theories in genetics— and seemed to contradict a central dogma of molecular biology, that information flows in one direction during transcription, from DNA out to proteins, in a cellular organism environment. John Cairns, Julie Overbaugh and Stephan Miller proposed a mechanism for what might be happening, that *"the cell could produce a highly variable set of mRNA molecules and then reverse-transcribe the one that made the best protein."* [6-10]

Another proven case of successful quantum biological adaptation exists in photosynthesis. Those who expect modern human technology to be faster and more efficient than natural processes might be surprised to learn that our best photovoltaic cells are only 20% efficient, compared with the 95% efficiency rates that photosynthesizing plants and bacteria regularly achieve for transforming sunlight into energy.

Anglo-Irish biologist JohnJoe McFadden believes many biological processes to be fundamentally quantum. MIT scientist Seth Lloyd, replied that this could easily be put to the test. Seth proposed that if, as JohnJoe suggested, plants were utilizing a quantum process such as a "quantum random walk" to transport energy from where photons hit the leaf to where the energy was stored, this could easily be proven or disproven, based on running a computer model. Much to Lloyd's surprise, the computer model matched the remarkable efficiencies observed in plant photosynthesis, proving JohnJoe McFadden's point that quantum processes really are happening in warm, wet, noisy biological environments. [6-11]

In decades past, scientists believed that excited electrons carried energy randomly through photosynthesizing plants, hopping from one molecule to the next. Modern measurements in quantum biology indicate electrons appear to be taking advantage of the fact that energy can move not just in material form, but as pure energy, too. Entire systems of molecules become entangled in photosynthesizing plants to allow formation of a coherent wave that tries out different pathways simultaneously, quickly determining the most efficient route. This quantum magic happens in each of a photosynthetic cell's millions of antenna proteins that are surprisingly efficient and robust at routing energy, with very little lost in transit.

University of Toronto biophysicist Greg Scholes shares a human analogy, to explain what these cells are doing, imagining that:

> *"… you have three ways of driving home through rush hour traffic. On any given day, you take only one. You don't know if the other routes would be quicker or slower. But in quantum mechanics, you can take all three of these routes simultaneously. You don't specify where you are until you arrive, so you always choose the quickest route."* [6-12]

Because quantum logic provides opportunities for exploring all options at the same time, natural systems can simply jump to the correct answer, with remarkable intuitive precision.

WIGNER'S FRIENDS AND WIGNER BUBBLES

In 1961, physicist Eugene Wigner proposed a thought experiment suggesting that observers of observers in quantum experiments could affect what is observed, and affect the outcome. [6-13] For example, a first observer might make an observation in the classic Schrodinger's cat experiment, where a cat is considered to be in a superposition of states of being both alive and dead inside of a closed box, until the box is opened and the truth is discovered. This initially shocking and seemingly preposterous idea of superposition of states arose from physicist Erwin Schrödinger's idea that a quantum particle mechanism (such as a radioactive isotope's random decay connected to a hammer poised to break open a vial of poison) would ensure that a cat is now involved in the seemingly impossible situation of joining the radioactive quantum isotope in also being in a quantum superposition of states. The first observer upon opening the box will find the system in a definite state of the cat either being dead, or very much alive.

The Wigner's Friend thought experiment raises the question of what an observer of the first observer will see, with questions regarding whether we can take another person's observations at face value, such that each person can have their own subjective reality, and whether we can trust our own observations from the past, since the past may be different from what was previously observed.

If you consider that right now you are in a quantum superposition, so your reading experience exists in several different quantum states simultaneously, then you can imagine that in one reality, you're reading this page, and in another you're thinking about these ideas and gazing out a window. If at this moment a second person is watching you, they could appreciate that you're either reading this page, or gazing out a window. You make one observation, deciding what you're doing, and someone observing you might see you doing another possible action.

Physicist Caslav Brukner conducted experiments acknowledging the Wigner's Friend paradox that we cannot assume the shared knowledge of other observers as our own, or even assume our past knowledge of the present, stating, *"... we have proved that one's own knowledge from the past cannot be used in the present, either."* Observing quantum systems is known to fundamentally change them. With respect to the Wigner's Friend paradox, Brukner's work shows that,

> *"After the friend's measurement has taken place, we are in a counterintuitive situation where Wigner describes the friend in quantum superposition of observing two different outcomes, while from the friend's perspective, a definite outcome must be perceived."* [6-14] [6-15]

Physicist Eric Cavalcanti coined the term *Wigner Bubble* to investigate implications of the Wigner's friend paradox, so that *"a coherent story can be told from the perspective of all agents involved regarding their own observations and betting commitments."* A probabilistic Quantum Baynesianism (QBist) perspective resolves the local friendliness aspect of the paradox by rejecting absoluteness of observed events. Cavalcanti insists this does not amount to solipsism or rejection of the relative existence of the friend's perspective, since probabilities are recognized with both Wigner and his friend sharing awareness of probabilities, and a final decision of what was observed coming into clarity based on their having placed a bet, which effectively "pops the bubble" resulting in quantum decoherence between observed outcomes, from Wigner's point of view. Cavalcanti describes such a personalist view of quantum states, where a quantum state is an encapsulation of your subjective degrees of belief as *Copernicanism.* [6-16]

NO SUCH THING AS OBJECTIVE REALITY

Reality appears to be subjective—not objective. I co-authored and presented a paper in 2017 on this topic with physicist George

Weissmann, *The Quantum Paradigm and Challenging the Objectivity Assumption*. We assert that the subject-object distinction has degenerated into an absolute split into separate realms, with scientists largely adopting the classical paradigm of "objective reality," and the idea that this can be studied independently from the subject pole of "the experiencer," expecting that this procedure yields a fundamental description of Nature. Typically, the subjective is eliminated altogether, and reduced to a presumed epiphenomenon of objective, measurable processes. By challenging the objectivity assumption—and leaving the question open—a coherent understanding of quantum nature and the quantum paradigm becomes possible. We then glimpse an amazing message from Nature that the basic components of objects—the particles, electrons, quarks, etc.—cannot be thought of as "self existent." In reality, they and all objects, are rather components of empirical reality—of experience. [6-17]   As physicist James Hopwood Jeans once said,

> *"The universe begins to look more and more like a great thought rather than a great machine."*

In one of the most remarkable news stories of 2019, scientists at universities in Edinburgh and Vienna demonstrated that two different observers could witness two completely different realities at the same place and time. Physicists at Heriot-Watt University in Edinburgh succeeded in bringing a classic Gedankenexperiment out of the realm of pure conjecture and into the real world of a physics laboratory. The thought experiment requires two people to observe one single photon– which is a quantum, or indivisible, unit of light. Quantum particles can behave as either particles, or as waves, settling into one state or the other (particle or wave) at the precise moment they are observed. All the rest of the time when a particle is not being observed by someone, it exists in a superposition of states in which it can be considered to be simultaneously both particle and wave. When a second person is unaware of the first person's observational measurement, this thought experiment proposes that the second person might be able to to confirm that the photon still exists in a quantum superposition (undecided) state. Scientists including Caslav Brukner at University of Vienna in Austria and Massimiliano Proietti at Heriot-Watt University in Edinburgh created an experimental apparatus involving lasers, beam splitters, and six entangled photons to be measured by equipment representing the role of the two observers. Results from this experiment showed that the two detection devices, although equally

146

reliable and accurate, observed two completely different realities—even though they made their observations at the same place and time. [6-6]

Nature provides us with natural quantum superpowers. We see evidence of quantum superpowers with bacteria that adaptively mutate within one generation to thrive in seemingly impossibly adverse circumstances, and where algae can utilize quantum coherence to attain optimal efficiencies for photosynthesis.

QUANTUM TIME

In a world where time itself is a bit of an enigma, it's amazing to see news of a recent mind-boggling discovery that time demonstrates quantum behavior. Physicists at the Stevens Institute of Technology, University of Vienna, and University of Queensland announced in 2019 that particles aren't the only ones capable of existing in a state of superposition–time can also exist in two or more states simultaneously. The scientists contemplated quantum temporal order, where no distinction exists as to whether one event caused another, or vice versa. They created a thought experiment examining Bell's theorem for temporal order, in which they examined answers to the question, *In a Quantum Future, Which Starship Destroys the Other?* Surprisingly, they found evidence of quantum properties of time, such that cause and effect can exist in both a forward and backward direction—thus giving the quantum starship captain the advantage. [6-18]

How could this be true? When envisioning an imaginary, hypothetical celestial space battle, we'd likely expect the rules of classical physics of very large objects, such as stars and planets, to determine the winner, but we would be wrong. As the researchers note, *"Sequences of events can become quantum mechanical."* What this means in layman's terms is that the Quantum Captain can win every time, with awareness of the gravitational time-slowdown when the classical physics Captain's ship is closer to a large astronomical body, such as a star. The physics rules that actually determine the winner in such a hypothetical space battle turn out to be quantum—not classical. Since causes can follow effects in quantum physics, it's possible for the quantum spaceship captain to win space battle games every time.

The double slit experiment has been called the most elegant experimental design in science. Physicist Thomas Young designed this experiment to investigate light's peculiar nature and ability to either behave like quantum particles, or quantum waves. The experiment consists of shining one photon of light at a time past two parallel narrow slits in a screen. The experimental apparatus consists of an observer, a particle emitter, a screen with double slits, and another screen. Amazingly, experimental results consistently indicate that photons of light can be fired through the apparatus one at a time, and depending on where an observational device is placed (such as inside one of the slits, or at the screen), the photons behave differently. If experimenters place a particle detector in one slit or the other, photons behave like particles; if they place no detectors in either slit, photons behave like waves. Physicist John Archibald Wheeler imagined several delayed choice thought experiments, where an apparatus that could be switched between a wave measuring apparatus, or particle measuring apparatus—AFTER a photon of light (that can travel as either a particle or a wave) has already traveled through most of the apparatus. Quantum delayed choice experiments are proving that future measurement decisions can influence events.

Quantum eraser experiments demonstrate how future choices of observational measurement perspective correspond with properties of quantum particles in the past, such that a decision made in the present can influence past events. The quantum eraser experiments are variations of Thomas Young's classic double slit experiment, first proposed in 1982 by Marlan O. Scully and Kai Druhl.

We can experience what retrocausality feels like any time we sense connections through space and time. For example, after my short interview conducted by Chris Anatra at the Idaho 2019 Mandela Effect conference was posted on YouTube, one viewer was inspired to comment, *"It's funny, as the interview almost was over I was thinking to myself, 'I wanna know why some people don't notice the Mandela Effect,' and a couple of seconds later you say, 'One more thing,' and bring up the subject."*

From my perspective, during my talk with Chris—in the past, long before the video was completed and viewed—I felt this question from the future and replied to it. This then provided an example of cause following effect. My answer to this question-from-the-future was, *"One more thing. If people are making fun of it, because a lot of people are starting to experience this, it's hard to talk to your friends and family. And there's a tendency sometimes that people say, 'Well, I know that's not true, because I'm an expert in this field.' And we've been talking about why sometimes the experts are not going to necessarily see the same changes that we do. And there's an explanation in physics for that, as well."*

This explanation refers to the Quantum Zeno Effect, where we can "lock in" a particular state of physical reality thanks to paying constant, steady attention to it, the way most subject matter experts do.

WE FOLLOW QUANTUM RULES

One of the top news stories of 2019 came from physicist Markus Arndt and his team at the University of Vienna in Austria, as they announced that "Even Huge Molecules Follow the Quantum World's Bizarre Rules." This finding challenges the long-standing assumption that there will always be one set of physics rules for larger objects, and the weird rules of quantum physics only for those tinier things "confined" to the Planck scale. [6-19] Arndt and colleagues observed quantum-like properties in very large molecules composed of 2,000 atoms, comparable in size to some proteins.

The quest to find where, exactly, a "seam" or boundary line of demarcation might exist between classical, relativistic and quantum "realms" meet has long fascinated physicists—particularly those inclined toward preferring a Bohrian Copenhagen collapse interpretation of quantum physics. Yet, as quantum logic and rules keep being verified at ever-more-massive scales, there is a distinct possibility that no such seam or dividing line exists.

Those of us experiencing Mandela Effects, reality shifts, and quantum jumps will be glad to hear that we are getting closer to seeing a respectable scientific explanation as to why these:

(1) can be expected to naturally occur,

(2) are an integral part of a recognized scientific model, and

(3) are necessary for life and reality as we know it to exist at all.

This is likely a good thing, because quantum physics keeps us mindful of possibilities. On some level, we're tuned in to all possible realities playing out around us in possible presents, futures, and pasts. We are capable of quantum jumping intuitively to optimal levels of efficiency.

EMBRACE SERENDIPITY

We can see how Nature embraces complexity at her very core, when looking at weather patterns. Complexity science and quantum physics describe Nature with concepts of: emergence, self-organization, and feedback—such that we more clearly see how events are necessarily subjective, incomplete, and impermanent. Thanks to a deep level of simplicity within Nature's hidden order, we can experience a sense of majesty in natural harmonious beauty. And we can have fun finding this beautiful hidden order when embracing the "random" with intuitive nonlinear living, as described in Chapter Two.

Author John Gribbin provides simple steps to create order out of randomness, following a simple, repetitive iteration. The Sierpinski gasket can be created by taking a blacked-out equilateral triangle that is drawn on paper or in a computer drawing program, and then taking the midpoints on each of the triangle's sides and create an upside-down triangle inside the first triangle that is white (or erased), in contrast to the surrounding black of the triangle it is contained within. This central "hole" now remains "empty," while an iterative process begins. By repeating this process for each of the three smaller triangles, and so on and so forth, we eventually begin to see the emergence of a pattern or shape of something akin to a triangular washer or gasket that is self-similar, and fractal. You can also create this same pattern utilizing just a pencil, paper, and an honest die—and begin a form of the "chaos game" in which a roll of the die randomly determines where to place a pencil mark that will ultimately generate the Sierpinski gasket after several hundred steps. [6-20]

UTILIZING THE QUANTUM ZENO EFFECT (QZE)

Most of us are familiar with those situations where the more we stare at a given situation, the less likely it seems we'll see anything interesting transpire. The old saying, *"A watched pot never boils,"* is more than just folklore, it is confirmed in the Quantum Zeno Effect (QZE). Alan Turing first mentioned the basic principle behind the Quantum Zeno Effect in the form of a paradox in 1954. In 1977, physicists Baidyanaith Misra and George Sudarshan hypothesized that if a quantum system is measured often enough, it's state can be locked in place, unable to progress. This hypothesis was proven in experiments involving laser-cooled ions trapped in electric and magnetic fields. In 2013, researchers demonstrated that objects as large as diamonds can exhibit the Quantum Zeno Effect. [6-21]

In 2014, a team of physicists led by Y.S. Patil at Cornell University demonstrated that rapid repetitive measurements can freeze a system in place. The potential implications of this are huge:

> *"The techniques demonstrated here… augur intriguing prospects of realizing novel many-body interactions such as a measurement-induced dynamic coupling between the internal, motional and topological states of a quantum many-particle system."* [6-22]

Fans of the science fiction TV show, *Dr. Who*, might remember the *Blink* episode featuring nefarious beings that look like statues of weeping angels. Dr. Who shared Quantum Zeno Effect tips with his friends with this warning,

> *"Fascinating race, the Weeping Angels. The only psychopaths in the universe to kill you nicely. No mess, no fuss, they just zap you into the past and let you live to death. The rest of your life used up and blown away in the blink of an eye. You die in the past, and in the present they consume the energy of all the days you might have had, all your stolen moments. They're creatures of the abstract. They live off potential energy. Don't blink. Don't even blink. Blink and you're dead. They are fast. Faster than you can believe. Don't turn your back, don't look away, and DON'T blink."*

While we likely won't have need to freeze Weeping Angels in place, there are times when we may feel a need to stop time, as I once experienced when walking through the train station in Lausanne, Switzerland, and seeing my daughter start to fall back head-first toward the hard marble floor. I needed her to be safe, and to somehow cover a distance of several steps to catch her safely in my arms. I heard and saw time slow to a stop, and was able to cover a great distance very quickly, just in time to safely catch my daughter. [6-23]

The Quantum Zeno Effect is far too useful and important to be cloistered away as the exclusive domain of physicists. Some scientists note that human perception can be influenced by it. Sudarshan, one of the original co-authors of the first 1977 paper about the Quantum Zeno Effect imagined a type of awareness in which:

> *"sensations, feelings, and insights are not neatly categorized into chains of thoughts, nor is there a step-by-step development of a logical-legal argument-to-conclusion. Instead, patterns appear, interweave, coexist; and sequencing is made inoperative. Conclusion, premises, feelings, and insights coexist in a manner defying temporal order."* [6-24][6-25]

The Quantum Zeno Effect may thus be helpful in accessing nonlinear intuitive ways of thriving in any situation, as well as handling emergencies. The Quantum Zeno Effect can also be useful for breaking long-standing habits and addictions, such as how to break cell phone addiction, as we illustrated in a short video we created, featuring physicist Henry Stapp talking about the quantum zeno mechanism for mental control over bodily action:

> *"The Quantum Zeno Effect says that the answers follow your questions—that by posing the questions fast enough, you can make the answers agree with what the questions are you ask."* [6-26]

An excellent application of the Quantum Zeno Effect involves focusing attentive awareness on what we most appreciate. Positive psychologists note that one of the most effective treatments for depression is to list a few things that we are grateful for that we had something to do with in the past 24 hours. This simple attention focusing activity has the power to lift some people out of very deep depressions. Another powerful way to positively focus attention is to do what some medical practitioners do who ask patients, *"On a scale of one to ten, what is your level of comfort,"* subsequently resulting in the observation that these patients require less painkiller medication and recover more quickly than people focusing on their pain.

SHYNESS EFFECT

The Shyness Effect got its name from physicist John Taylor in his book, *Superminds: A Scientist Looks at the Paranormal.* Taylor writes,

> *"One curious feature of the bending process is that it appears to go in brief steps; a spoon or fork can bend through many degrees in a fraction of a second. This often happens when the observer's attention has shifted from the object he is trying to bend. Indeed this feature of bending not happening when the object is being watched—'the shyness effect'—is very common. It seems to be correlated with the presence of sceptics or others who have a poor relationship with the subject."* [6-27]

My first encounter with the shyness effect occurred when I was a young girl of about five years of age, looking out our living room window into the backyard, watching the rain and noticing that the rain seemed to directly respond to my thoughts, such that when I thought, "Stop rain," the rain would immediately stop. And when I thought, "Start rain," the rain would start again. When I ran to show this to my mom, try as I might, I could not get this demonstration to work.

When I first met Uri Geller in San Francisco in 1999, I was intrigued to witness something similar happen to him in one of his psi demonstrations. He'd asked a volunteer to "think of a color," and at some point, Geller told us he was picking up interference from someone sending conflicting telepathic information. A man standing in the back of the room admitted that yes, he had been telepathically sending a conflicting color of his own. I appreciate the way this "shyness effect" principle provides a beautiful compliment to the Quantum Zeno Effect (QZE), with an indication of the importance of the focus of attention and intention of those around us.

NATURAL QUANTUM EXPERIENCES

Nature gifts us with amazing quantum abilities, so we can acknowledge that the paranormal is normal, since it's perfectly natural. Table 6-1 includes some key principles from quantum physics, listed alongside natural experiences that many of us have had.

We get a feeling of the interconnection of quantum entanglement in dreams, intuition, muscle testing, and dowsing. We get a feeling for the togetherness of quantum coherence in shared dreams, empathy, synchronicity, telepathy and ESP. We sense quantum superposition in dreams, daydreams, embodied cognition, the placebo effect, and spontaneous remissions. We can encounter quantum tunneling in dreams, daydreams, sensory perceptions, and if we teleport to safety. We can witness quantum steering via psychokinesis.

Table 6-1:

Quantum Terms & Natural Experiences

Scientific Term	Natural Experiences
Quantum Entanglement	Dowsing, Dreams, Daydreams, Intuition, Muscle Testing
Quantum Coherence	Dreams, Daydreams, Empathy, Synchronicity, Telepathy, ESP
Quantum Superposition	Dreams, Daydreams, Embodied Cognition, Placebo Effect, Spontaneous Remissions
Quantum Tunneling	Dreams, Daydreams, Sensory Perceptions, Teleportation to Safety
Quantum EPR Steering	Psychokinesis, psi influence

MANDELA EFFECT IN THE QUANTUM AGE

The Quantum Age invites us to embrace uncertainty, recognize interconnectedness, and feel energized about experiencing a better life. Quantum phenomena are so intrinsically part of Nature that they can no longer be presumed to be solely relegated to the dominion of quantum mechanics. For example, we now know that human cognition and decision making are quantum processes. We know that plants photosynthesize according to quantum principles. These quantum principles are becoming so apparent that at some point, we will need to acknowledge that quantum logic and phenomena can prevail at all levels of scale and size. Table 6-2 maps out our Quantum Superpowers, or how we can relate some common Mandela Effect phenomena to their quantum physics counterparts.

Through **quantum entanglement**, we can appreciate how entire histories and backstories sometimes change after experiencing Mandela Effects shared in "Wigner bubbles" or groups of people noticing similar effects. Thanks to entanglement, when observing a given Mandela Effect, all associated aspects of that Mandela Effect typically change simultaneously, via what Albert Einstein called, "spooky action at a distance." Those familiar with the "Law of Attraction" may recognize the importance of aligning oneself with desired outcomes; this can be viewed as proactively quantum entangling oneself with an energetic match. You can select your preferred possible realities by reaching out with joyful emotions and a confident sense of what you envision, love, and need—becoming coherent with that state of being.

Through **quantum coherence**, we better understand synchronicity and meaningful coincidence, experiencing the ways ideas and information can instantly be received by many at the same time, like returning to a larger Wigner bubble of shared mainstream thoughts. We can thus consider that when we observe ourselves and others "Taking the Download," we may be meeting the agreed-upon collective consensus reality—much the way a starling separated from its flock can rejoin the others, all moving with quantum coherence, as if they are one.

Quantum entanglement and quantum coherence can be witnessed in Mandela Effects when observing how when one thing changes, all related facts also change at that same time. For example, when I was a little girl, I brought my Peanuts metal lunchbox to school every day, and spent hours gazing at it's cartoon artwork and the signature of Peanuts creator, Shultz. Official Peanuts products have now always been created by "Charles Shulz" with no letter "t" in the name anywhere.

Table 6-2:

Quantum Superpowers

Quantum Phenomena	Mandela Effect Phenomena
Quantum Entanglement	The backstory changes after Mandela Effects occur, in Wigner bubbles and groups of collective consciousness
Quantum Coherence	Share historical facts with those in same Wigner bubbles & groups
Quantum Superposition	Choose between possible parallel realities revealed by ME flip-flops
Quantum Jumps & Quantum EPR Steering	Focus attention & intention with energized, loving, conviction
Delayed Choice in Participatory Universe	Receive insights from Future Memory and déjà vu
Quantum Eraser	Practice optimal detached observation, unobservation, and reobservation, to cut through "noise"
Quantum Teleportation	Travel farther in less time, and teleport
Quantum Tunneling	Walk through walls, move through objects, objects appear & reappear
Quantum Random Walks	Follow the random, intuitive guidance
Quantum Zeno Effect (QZE)	Lock in preferred realities (such as "hear what you want to hear")
Closed Time-like Curves	Time loops & time shifts

Through **quantum superposition** we can glean insights into how we may witness spontaneous remissions from disease occurring when people encounter higher and more loving states of consciousness and mind, making a jump to an optimal reality. We sometimes glimpse adjacent possible realities through "flip-flops" back and forth between two or more possibilities. We might observe, for example, that our dog has cataracts, while in another reality we know our dog does not have cataracts. In noticing such flip-flops and remembering both realities, we are sensing a superposition of states.

Through **quantum jumps and quantum steering**, we can gain experience of making timeline edits that can provide individual and group benefits. We can experience optimal reality shifts when energized with internal energy/ki with deep loving connection to a given reality, and total conviction that's the reality we choose.

Through **delayed choice in a participatory universe**, we can receive future memories guiding us to make optimal choices, by inviting and encouraging inspiration from our future selves. We can do that by viewing every important creation, decision, and plan as if we are seeing it through the eyes of our future self looking back on it from a few years from now, as if it's done.

Through the **quantum eraser**, we can focus on entanglements—with what and who we truly love and care most about—to find deeper meaning and messages amidst seemingly chaotic interference and noise. Quantum erasers can help us stay detached, and unobserve that which we'd rather not experience. We can observe, unobserve, and reobserve in a calm state of loving detachment, thus focusing attention on what we genuinely and truly need, love, and desire.

Through **quantum teleportation**, we see how we can sometimes travel farther in less time—sometimes instantly, and how we and various objects can teleport.

Through **quantum tunneling**, we sometimes notice objects appear or reappear, experience moving through objects, or walk through walls.

Through **quantum random walks** we can "follow the random," and benefit from trusting intuitive guidance, living with less stress.

Through the **quantum zeno effect (QZE)**, we can lock in a preferred reality through sufficiently frequent observations of the desired state. For example, we can sometimes hear what we want to hear, or see what we want to see, and experience those realities.

Through **closed time-like curves (CTC)**, we can witness changes of the experience of time, and looping of events leading to a number of different types of possible time shifts, such as conversing with future selves.

When appreciating the many ways that Nature blesses us with quantum superpowers, we can open our minds to new ways of rising above whatever situations may seem to be arising in our lives. When we know that history can change, we can play a more fully actuated role of being the causal agents we've been hoping for. By knowing that quantum physics forbids a single history, we can invite the changes we wish to see in the world. We can thus harness Mandela Effect technologies of quantum superpowers and skills, while becoming multidimensional conscious agents in this new Quantum Age.

MANDELA EFFECT
SCIENTIFIC PROCESSES

This list of scientific terms and ideas is provided to help improve understanding of the Mandela Effect. These ideas can help answer our questions about how the Mandela Effect works, what it is, and how we might some day come closer to scientifically validating it, as well as optimizing how we can work with natural quantum principles in our everyday lives.

> **Alternate Histories** are the idea that many sequences of past events are possible. This concept is intrinsically part of quantum physics. Physicist Stephen Hawking stated that *"Quantum mechanics forbids a single history,"* suggesting that the universe has no singular beginning, but rather came about in just about every way imaginable. [6-28] [6-29]

> **Anti Zeno Effect** is the principle by which a quantum system that has been "locked" into a given state can quickly break free, the opposite of the Quantum Zeno Effect. (see: Quantum Zeno Effect)

> **Bicausality** suggests the idea that the present is capable of influencing the past, in addition to the idea that events in the past affect the present. Physicist and retrocausality conference organizer Daniel Sheehan points out, *"To say that it's impossible for the future to influence the past is to deny half of the predictions of the laws of physics."* (see: Retrocausality)

Black Holes are regions of spacetime that are so massive that they attract all matter and electromagnetic radiation nearby. They might not have singularities at their center, but draw matter inside and send it across space to some future time, thereby providing a means to support time travel.

Bohmian Dialogue is a type of conversation in a group of people who talk and deeply listen to one another, without hierarchy. Physicist David Bohm said, *"Dialogue is really aimed at going into the whole thought process and changing the way the thought process occurs collectively. We haven't really paid much attention to thought as a process. We have engaged in thoughts, but we have only paid attention to the content, not to the process."* [6-30]

Closed Time-like Curves (CTC) might provide the means for traveling backwards in time, or to other adjacent possible universes through *backward time steps*. Physicist Fred Alan Wolf points out that classical *mixed states* arise when observations occur in closed time-like curves, possibly representing the action of a conscious mind. [6-31]

Coherence in quantum physics is said to exist when there exists a definite phase relation between different states. Coherence can be seen when starling birds move "as one" with every bird turning exactly at the same time, moving with perfect synchrony, as in a quantum fluid such as liquid helium. (see: Decoherence)

Complexity science studies dynamic, multi-dimensional, nonlinear systems. Qualities of complex systems evident in quantum phenomena include: self-organization, sensitivity to initial conditions, a fractal quality, emergence, feedback, and steerability. [6-32]

Copenhagen Interpretation of quantum mechanics was proposed by physicists Niels Bohr and Werner Heisenberg. It considers there to be two realms: a macroscopic, classical realm of measurements that is governed by Newtonian physics; and a microscopic, quantum realm of tiny particles/waves governed by the Schrödinger equation, such that choice of observational perspective and method appears to collapse the wave function into a particular state.

Copernicanism is a term coined by physicist Eric Cavalcanti to describe a personalist view of quantum states, which are encapsulations of your subjective degrees of belief. [6-16]

Decoherence is the loss of quantum coherence associated with a system, such that it no longer demonstrates quantum characteristics. Particles that have previously been entangled could become connected and go separate ways, such that a particle in superposition would split into two versions of itself. Decoherence can thus be thought of as an opposite process to superposition. (see: Coherence)

Delayed Choice is the idea proposed by physicist John Archibald Wheeler postulating that decisions made in the future can affect the present. This idea was proven to be true in versions of the double slit experiment, including some demonstrations of future choices in systems that include both massive particles and massless photons. [6-33]

Double Slit Experiment is a famous experiment providing mind-boggling insights into the behavior of quantum "particles." The basic double slit experiment is arranged so single photons are fired one at a time through a pair of slits before landing on a screen. When photons act like waves, characteristic peaks and valleys of wave interference can be observed. The determining factor of whether the photons traverse the apparatus as waves or particles is decided by how observational devices are positioned. If experimenters place a particle detector in one or the other slit, the photons behave like particles; if they do not place a detector in either of the two slits, the photons behave like waves. Physicist John Archibald Wheeler famously posited a group of delayed choice thought experiments, where he envisioned an apparatus that could be switched between a wave measuring apparatus, or particle measuring apparatus—AFTER a photon of light (that can travel as either a particle or a wave) has already traveled through most of the apparatus.

Einstein-Rosen Bridges are theoretical structures linking various points in spacetime, like a tunnel with different starting and ending points in space, in time, or both.
(see: Wormhole)

Embodied Cognition is the process by which physically making various bodily movements can bring about the emotions associated with such physical moves. Examples include smiling and subsequently feeling more joy, and stretching arms up and out and subsequently feeling more confident and less stressed.

Entanglement is a physical phenomenon occurring when a pair or grouping of particles move together as one, even when separated by great distances. Albert Einstein referred to this as *spooky action at a distance*. Entanglement provides animals and plants with instantaneous telepathic communication. Our plants and pets can know the exact moment that we are starting to head home when we make that decision, as documented in Cleve Backster's book, *Primary Perception*, and Rupert Sheldrake's book, *Dogs That Know When Their Owners are Coming Home*. [6-34] [6-35] Theoretical neuroscientist Walter Jackson Freeman III recognized the importance of considering the way societies of minds and brains work together and are more interconnected than scientists typically acknowledge. [6-36]

EPR Steering is an idea introduced by physicist Erwin Schrödinger to describe entanglement that allows experimenters to non-locally affect, or steer, another system's states through local measurements, as a generalization of the Einstein-Podolsky-Rosen (EPR) paradox. (see: Quantum Steering)

Explicate Order is part of the Holographic Interpretation of quantum mechanics proposed by physicist David Bohm, where mind and matter are related projections. (see: Implicate Order)

Goldilocks Zone is an attentional area where reality shifts and Mandela Effects can be observed without either being completely locked in (as with QZE), or completely missed due to no awareness by an observer of those events (shyness effect).

Holographic Interpretation is an interpretation of quantum mechanics proposed by physicist David Bohm, in which a holistic principle of undivided wholeness is combined with the idea that everything is in a state of process or becoming (*universal flux*).

Implicate Order is part of the Holographic Interpretation of quantum physics proposed by physicist David Bohm. (see: Explicate Order)

It from Bit is the idea proposed by physicist John Archibald Wheeler that every material thing fundamentally consists of pure immaterial informational "bits."

Loop Quantum Gravity is postulated by physicist Carlo Rovelli, as a "theory of everything" that quantizes General Relativity, with quanta (indivisible building blocks) of gravity being linked by loops. Space could thus be a collection of non-particles, woven by gravity quanta. The quantum version of Faraday's lines of force can thus weave and create space, like a three-dimensional mesh of interlinked loops.

Many World Interpretations of quantum mechanics have been proposed by physicists including Hugh Everett III, David Deutsch, and others, based on the premise of the reality of parallel worlds.

Measurement Problem is related to the Observer, and is considered a problem in physics, because there appears to be no singular fixed reality existing when nobody is looking. Intriguingly, there can exist layers upon layers of observers—such that it is possible for someone observing an observer to influence events—such as someone observing someone observing Schrödinger's cat inside the box with the poison.

Nocebo Effect is a special case and reversed example of the Placebo Effect, in which someone experiences harm or injury, based purely on belief. The Latin term "nocebo" translates to "I harm." (see: Placebo Effect)

No Go Theorem by physicists Frauchiger and Renner in 2018 points out that *"quantum theory cannot consistently describe the use of itself,"* describing three assumptions that cannot all be correct: (1) absoluteness of observed events, (2) no super-determinism, and (3) locality. [6-37]

Objectivity is the idea that there is one "true perspective" by which everything can be accurately described. There appears to be no such thing as "objectivity," since time is necessarily subjective, as are observations.

Observation plays an essential role in many quantum mechanics theories, where selection of measurement method affects what is (or was) observed. There is a special meaning to the word "Observer" in quantum physics, in which selection of a method of measurement of a quantum system influences what is subsequently observed. Methods of measurement include people at some point, so we play the role of Observer when we choose what we look at and how we look at the world. This aspect of quantum physics is sometimes referred to as "the observer effect" and "the measurement problem."

Observer Effect is the idea that the act of observation of an event affects what is being observed. With respect to psychology, this can take the form of "observer bias" with "observer expectancy effect"; with respect to physics, the selection of the type of observation to be made plays a vital role in determining the outcome that is observed.

Participatory Universe is the concept described by physicist John Archibald Wheeler in which conscious observers play a role in making quantum potentials real, with ordinary matter and radiation playing dominant roles. Wheeler explained: *"It from bit symbolizes the idea that every item of the physical world has at bottom—at a very deep bottom, in most instances—an immaterial source and explanation; that what we call reality arises in the last analysis from the posing of yes-no questions and the registering of equipment-evoked responses; in short, that all things physical are information-theoretic in origin and this is a participatory universe."*

Perspectival Objectivity was proposed by Peter W. Evans to accommodate subjectivity in quantum mechanics, based on a notion of observer-independence indexed to an agent perspective. [6-38]

Placebo Effect is the remarkable efficacy of a variety of treatments to bring about curative effects, despite having no known attributes to account for their healing powers. The term *placebo* means "I shall please" in Latin. The opposite of placebo is known as *Nocebo*. (see: Nocebo Effect)

Quantum Baynesianism (QBism) is an agent-centered interpretation of quantum theory that considers theoretical quantum physics from the foundational idea that probabilities are subjective degrees of belief. QBism has roots in personalist Baynesianism probability theory, emphasizing the observation that quantum phenomena arise in states of uncertainty (i.e. the realm of probabilities), and never in physical mechanical configurations. [6-39]

Quantum Biology is the study of the role of quantum effects (such as quantum random walks, quantum tunneling, and quantum teleportation) in biological systems, as described by quantum biologist JohnJoe McFadden. [6-40]

Quantum Cognition appears to be the way humans naturally think, with awareness of the significance of relationships and interconnections. The order of choices provided can thus influence how people will respond, with evidence that human cognitive functioning can be modeled by quantum operations,

162

such as described by Jerome Busemeyer and Peter Bruza. (see: Quantum Decision-Making)

Quantum Computing was first proposed in the 1980's, based on quantum bits or *qubits* containing information in all possible states, via quantum superposition of states. Entangled qubits can compute optimal solutions for some of the most complex problems, so quantum computing is expected to revolutionize: the stock market, freight transportation, information security, weather forecasting, and trend analysis.

Quantum Decision Making models can accurately predict human probability judgments, as well as the way people's answers tend to change depending on the sequence of questions being asked, various types of subconscious reasoning, and the way words and meanings are grouped. Cognitive scientist Zheng Wang and colleagues appreciate quantum theory's potential to build better models of human cognition. [6-41] (see: Quantum Cognition)

Quantum Erasers involve the ability to both observe and "unobserve." In quantum mechanics, the idea of a quantum eraser was proposed in 1982 by Marlan Scully and Kai Druhl, such that wave-like behavior could be restored by removing—or erasing—path information. One's ability to observe, unobserve, and reobserve is akin to mastering the art of being sufficiently detached from what seems to be happening according to sensory perceptions, while staying primarily focused on the underlying energetic realm of possibilities; "being in the world, but not of it."

Quantum Immortality refers to the theory that none of us can actually die, sometimes adding that we are already dead. This "quantum suicide" idea was first proposed as a thought experiment by Hans Moravec in 1987 and independently developed by Bruno Marchal in 1988. (see: Quantum Suicide)

Quantum Jumps can occur with jumps via "wormholes" or through proverbial "rabbit holes." Entangled black holes provide pathways through a holographic multiverse. If a wormhole can form between a black hole and its emitted photons, these Einstein-Rosen bridges could be viewed as entanglement between black holes, as Leonard Susskind and Juan Maldacena point out. MIT's Julian Sonner shows that the creation of a pair of entangled quarks gives rise to a wormhole connecting the two of them.

Quantum Logic differs from Classical Boolean logic, whose binary true/false choices give us a narrow, special case view of a world quite different from Nature. Quantum logic embraces the kind of four-fold logic evident in the natural world, in which the categories of true-and-false and not-true-not-false are the largest groups.

Quantum Nonlocality is the way that quantum systems don't allow for an interpretation in terms of a local realistic theory, and instead allow for entangled parts of a quantum system to move together simultaneously, no matter how far apart they may be. (see: Spooky Action at a Distance)

Quantum Potential is part of the de Broglie-Bohm formulation of quantum mechanics, as an information potential which acts on a quantum particle.

Quantum Random Walks are a type of randomly prescribed steps that tend to lead directly to very efficient solutions in quantum systems. Plants have been found to photosynthesize according to quantum random walks, ensuring optimal efficiency of photosynthesis. In contrast to classical random walks, quantum random walks enjoy exponentially fast speeds and efficiencies.

Quantum Steering suggests how observers might influence experimental outcomes. The concept of EPR-steering was first proposed by physicist Erwin Schrödinger as a thought experiment in which remote observers of a quantum system share entangled particles. In such a scenario, one observer might then be able to direct—or steer—the state of another's system when performing a measurement on their system. In hypothetical experimental quantum steering situations, one individual called "Alice" might influence the observations of a quantum system by another observer, "Bob." Related areas of research involve determining if this kind of influence is one-sided, whether Bob can tell when he is being influenced, whether such influence can occur even if Alice is known to Bob as "untrustworthy," and how strong such influence might be. [6-42] [6-43] (see: EPR Steering)

Quantum Suicide was first proposed as a hypothetical scenario to demonstrate the Schrödinger's cat thought experiment, from the cat's point of view, by Hans Moravec in 1987 and this idea was further developed by Bruno Marchal in 1988. (see: Quantum Immortality)

Quantum Teleportation refers to a process by which two distant particles are entangled, and a third particle instantly teleports its state to the two entangled particles. Quantum teleportation can feel life-saving for those who have reported instantly being on the other side of oncoming vehicles, or have otherwise been teleported to safety.

Quantum Tunneling is a quantum phenomenon in which a subatomic particle can disappear on one side of a physical barrier and appear on the other. It seems likely that our sense of smell is made possible thanks to quantum tunneling, as first proposed by Luca Turin. [6-44]

Quantum Zeno Effect (QZE) holds a system in place with frequent observations or measurements, such that quantum "watched pots" will never boil. QZE helps explain why subject matter experts are less likely to notice Mandela Effects than those not so closely connected with a given area. For example, many medical professionals who constantly work with certain parts of the body may not see any anatomical changes, due to their constant checking in those areas—so for them, there is no change. (see: Anti Zeno Effect)

Retrocausality is a principle of quantum physics from time symmetry that presents us with ideas of cause and effect in which the future can influence the present. Physicist and retrocausality conference organizer Daniel Sheehan points out, *"To say that it's impossible for the future to influence the past is to deny half of the predictions of the laws of physics."* (see: Bicausality)

Schrödinger's Cat thought experiment is a seemingly paradoxical illustration of physicist Erwin Schrödinger's concern that something as large as a cat could be placed in a superposition of states inside a closed quantum system. The hypothetical cat is placed inside a sealed box with a flask of poison and a quantum random triggering device that might at any moment break the poison flask and kill the cat. The paradox is that while the cat is inside the box and before it is observed, it can be considered to be in a superposition of states —being both dead and alive.

Shyness Effect happens when metaphysical phenomena is less likely to occur when intensely observed (such as with QZE), or when expected to be repeated with mechanistic predictability, rather than via natural inspiration. Quantum systems can be provided freedom from being inadvertently "locked in place" by being granted private times of inobservance.

Solipsism is the view or theory that the self is all that can be known to exist.

Spooky Action at a Distance was the nickname given to quantum non-locality by physicist Albert Einstein. (see: Quantum Nonlocality)

Spontaneous Remission is a name given to a case of instantaneous or sudden absence of a disease or injury condition with no known healing causal agent. The Institute of Noetic Sciences (IONS) has documented numerous such cases.

Stochastic Resonance is a natural phenomenon that demonstrates how noise aids coherence—such that a quiet message can be better received amidst background noise.

Subjective Observations show seemingly contradictory, yet trustworthy, reports of events that happened at the same place and time. Caslav Brukner and Massimiliano Proietti created an experimental apparatus involving lasers, beam splitters, and six photons to be measured by various equipment representing the role of the two observers. Preliminary results indicate that our assumption of shared objective reality may be inaccurate. [6-45]

Superposition of States involves being in more than one state at the same time. Superposition helps explain how people may have memories both of how something was before it changed, as well as memories of how it was after a reality shift—and also why people don't all experience the same Mandela Effects. Superposition is acknowledged by a majority of physicists to occur at all levels of reality—and not just "in the quantum realm." Mandela Effect flip-flops are evident when we notice something that is first one way—and then another.

Time Travel might be possible through harnessing black holes or wormholes. David Deutsch theorized in 1991 that "time travel paradoxes" (such as the grandfather paradox) created by closed time-like curves could be avoided on the quantum scale, thanks to quantum particles following fuzzy rules of probability.

Transactional Interpretation of quantum mechanics was proposed by physicist John Cramer to consider both forward-in-time (retarded) and backward-in-time (advanced) wave functions, interacting in a kind of "quantum handshake." [6-46]

Uncertainty Principle does not allow us to know both an object's velocity and it's position at the same time; we can know one or the other, but not both. Heisenberg proposed that quanta must be described as probability waves while they travel, and as particles (matter) when they are viewed. We cannot accurately make definitive predictions, other than establishing a sense of statistical probability for certain outcomes.

Wave-Particle Duality is a property of matter, such that all matter exhibits both wave and particle properties.

Wholeness is part of the Holographic Interpretation of quantum mechanics as proposed by David Bohm.

Wigner Bubble is Eric Cavalcanti's idea that we reject the assumption of Absoluteness of observed events, so not all observers will agree about what transpired: *"In this sense, the events that are definite for the friend but not for Wigner could be said to not be in 'Wigner's spacetime,' but occurring in a 'Wigner Bubble.'"* [6-16]

Wigner's Friend is a thought experiment proposed by Eugene Wigner in 1961, and further developed by David Deutsch in 1985, where one observer performs a quantum measurement on a physical system, while another observer observes the first observer. This experiment reflects the seeming incompatibility between deterministic, continuous time evolution of the state of a quantum system, and probabilistic, discontinuous collapse of a system when measured.

Wormholes are also known as Einstein-Rosen bridges; they are speculative structures linking various points in spacetime, like a tunnel with two ends at two different places in space, or two different points in time, or both. (see: Einstein-Rosen Bridges)

* * * * * *

The next chapter, *Mandela Effect Experiencers,* investigates who the Mandela Effect-affected are, what experiencers have in common, and how we can best relate to ever-changing pasts.

~~~~~~

EXERCISE:
WHAT FASCINATES YOU MOST?

This is a truly marvelous time to be alive, when we have ideas and language to describe intriguing aspects of quantum physics and the Mandela Effect. Some of these may already feel familiar to you, and others may seem more exotic. Simply by noticing one of these pairs of quantum-Mandela Effect "superpowers" in Table 6-2 that you find most fascinating, you can naturally learn some associated steps that work best for who you are right here and now.

~~~~~~

EXERCISE:
PLAY WITH THE QUANTUM ERASER

The Quantum Eraser works more through re-routing what we're connecting to, rather than literally "erasing" past or possible events. A key benefit of the Quantum Eraser is that we can minimize interference—of the sort that might distract us from time-to-time when our attention is best focused on the horizon of our long-term goals, rather than on the next pot-hole in the road, or what seems to be happening. Think of a goal or desired outcome that you have felt thwarted from attaining, that you'd truly love to achieve. Now, instead of ruminating on habitual thought patterns associated with this idea, imagine all the good possibilities as well as past, present, and future supporters who'd love for you to succeed. This can include friends you've not yet met.

~~~~~~

EXERCISE:
PLAY WITH THE QUANTUM ZENO EFFECT

Think of one of your favorite regularly occurring moments in daily life that you'd like to savor even more. Imagine that you can slow down time to appreciate every sensory perception during these cherished moments. Imagine you can master the art of "locking in" favorite moments, and experiencing time slowing down when you are having the most fun, so these treasured golden moments can be expanded, revered, and appreciated.

~~~~~~

~~~~~~

EXERCISE:
PLAY WITH THE SHYNESS EFFECT

Boost your manifesting game by having faith that "God's got this" and "God does the heavy lifting" while thinking of or writing down intentions for your day. Then, simply release these wishes, the way you might blow the seeds off a dandelion, or blow out birthday candles on a cake, and let them go, with full faith and trust that Creator/Source/God/Eternity is doing the heavy lifting, and all you need to do is just show up and do the little things the best you can.

~~~~~~

*"No single person can liberate a country.
You can only liberate a country
if you act as a collective."*

—Nelson Mandela

Chapter 7

MANDELA EFFECT EXPERIENCERS

In 1998, I sought out some kind of community that I knew must exist, searching for kindred spirits on the new, self-organizing internet with the term, *reality shifts*. That phrase seemed to best capture the nature of a phenomenon I was observing where it seemed clear that reality was changing, anywhere and everywhere. I found some of the earliest community members through the ThirdAge website in the late 1990s, long before more current social media sites such as Reddit, YouTube, X, and Facebook came along.

When I created an online discussion group for people to explore reality shifts and what we now know as the Mandela Effect on the ThirdAge website in 1998, those of us who were observing differences between what we remembered and official historical versions of reality shared respectful appreciation for one another's unique experiences. A few years later in February 2001, we created an online community for conversations about this subject in a realityshifters discussion list at yahoo, providing a safe haven for people to describe their first-hand experiences, where they could be respectfully appreciated.

In April 2000, I surveyed 395 participants to investigate which kinds of reality shifts were most commonly experienced. People were encouraged to complete this survey regardless of whether they believed in the phenomenon of reality shifts, or noticed they had ever experienced reality shifts. An overwhelming 95% majority of survey respondents reported noticing synchronicity and coincidence in their daily life; with 86% noticing that time seemed to slow down, stop, or speed up; and 78% often finding parking spaces when and where they most needed them. Less commonly reported was having bent spoons or other metal without forcing the objects to bend (8%), and not yet having ever witnessed any reality shifts (6%). Of those who noticed reality shifts, the most common emotional reaction to witnessing the shifts were feelings of curiosity (62%) and excitement (45%), followed by feeling awestruck (37%), happy (33%), and confused (26%). 59% of respondents noticed reality shifts had changed their attitudes, 55% felt that observing reality shifts had changed their beliefs, and 45% felt that both their attitudes and beliefs had changed as a result of witnessing reality shifts. [7-1]

Over the years, similar communities and conversations sprung up across the internet. In 2010, Paranormal researcher Fiona Broome created a conversational forum at her mandelaeffect.com website, and other groups formed on reddit, Facebook, and YouTube. Each group and community has it's own distinctive character, attracting those who resonate with its guiding goals, beliefs, and mindset. While some focus on how the Mandela Effect provides us with evidence of humanity's evolution of consciousness, others focus on other areas—such as concerns regarding activities at CERN's Large Hadron Collider. Mandela Effect communities on reddit provide the ability to remain anonymous, which many people appreciate when first publicly exploring these pioneering ideas.

WHY US? WHAT DO WE ALL HAVE IN COMMON?

American chiropractor Dr. Tarrin Lupo creates videos about food and health. In January 2017, Dr. Lupo posted a video with a survey entitled, "Why us? What do we all have in common?" Dr. Lupo elaborated, *"Why are some people experiencing the Mandela Effect and others are not? Do we all have something in common that is making us susceptible to the effect? Let's figure it out together. I explore some common characteristics that people are talking about and conduct a poll, to try to find some answers."* [7-2]

By December 2019, Dr. Lupo's video had received over 27,000 views, and received answers to five questions about such things as spiritual beliefs and diet. About two-thirds of Mandela Effect-affected respondents indicated they were spiritual (67%), 16% indicated they were religious, 10% agnostic, and 4% atheist. Perhaps the most noteworthy quality surveyed had to do with empathy; 86% of Mandela Effect-affected respondents replied they had always considered themselves to be an Empath, with 3% saying they had developed empathic abilities after experiencing the Mandela Effect, and 9% saying they had never been an Empath. Dr. Lupo's third question had to do with cognitive functioning: 17% of experiencers had Eidetic (photographic) memory, 15% indicated they had Obsessive-Compulsive Disorder, 4% said they had learning disabilities, 27% indicated they had more than one of these, and 35% had none of those. Dr. Lupo's fourth question inquired whether these experiencers had always distrusted government or authority. 68% indicated that was the case, 26% indicated they had developed distrust of the government over the past five years, 2% indicated they had developed distrust only after experiencing the Mandela Effect, and 2% indicated they had never distrusted the government and authority. Dr. Lupo's fifth question had to do with what kind of diet experiencers ate: 40% ate the standard American junk food diet, 42% were on the Carnivore-Paleo type diet, 12% were Vegetarians, and 3% were Vegan. There was an open-ended sixth question that Dr. Lupo asked, regarding whether experiencers had a vivid memory of almost dying or having a close call with death, and asking people to please indicate if this was the case in the comments for this video. There was a relatively small percentage of comments indicating people had such memories.

WHAT IS THE MANDELA EFFECT PERSONALITY?

In 2018, I surveyed 470 people, to discover what correlations existed, if any, between Myers-Briggs personality types and Mandela Effect experiencers. What I found was astonishing: while Myers-Briggs "Intuitive Feelers" account for only 14% of the overall population, they are more likely than other personality types to observe Mandela Effects, comprising 65%—almost two-thirds—of the Mandela Effect Experiencers. Why would some of the rarest Myers-Briggs personality types be more likely to experience Mandela Effects and reality shifts?

The Myers-Briggs Type Indicator (MBTI) personality inventory was developed by mother-daughter team Katharine Cook Briggs and her daughter, Isabel Briggs Myers. Myers and Briggs applied some of Swiss psychologist Carl Jung's ideas and insights to recognizing human personality characteristics. According to Carl Jung's theory of psychological types, people can be characterized according to tendencies toward: Extroversion or Introversion, Sensing or Intuition, Thinking or Feeling, and Judging or Perceiving. These result in 16 personality types. [7-3]

The key principle underlying the foundation of the Myers-Briggs personality test is that much seemingly random variation in peoples' behavior can be recognized as actually being orderly and consistent, since peoples' behavior is driven by intrinsic differences in individual personal preferences and experiences of the world. The Myers-Briggs questionnaire investigates individual inclinations between such things as preferring to focus on the internal (Introvert) or external (Extrovert) world. The resulting sixteen personality types present us with an overview of our uniquely personal perspectives. These personality types do not indicate nor represent character, and there is no "best type." Thanks to pursuing true interests that suit our natural way of being, we can feel inspired while sharing our talents with one another, as we gravitate toward preferring to be teachers, chefs, engineers, nurses, scientists, artists, or story-tellers.

Within the sixteen Myers-Briggs personality types, another pattern can be seen in which some people have "NF" at the core of their category. This "NF" stands for "Intuitive Feeling" and INtuitive Feelers (NFs) are one of the basic temperance types, the others being: INtuitive Thinkers (NTs), Sensate Judgers (SJs), and Sensate Perceivers (SPs). David Keirsey identified 16 personality types based on temperaments noted by the Greek philosopher, Plato: Idealist (noetic), Rational (dianoetic), Artisan (iconic), and Guardian (pistic) and correlated these four temperaments into two roles, each with two types of role variants. [7-4]

The Myers-Briggs Foundation is frequently asked how common each type is, and shares the percentages found in the USA from 1972 to 2002 as being: 13.8% ISFJ, 12.3% ESFJ, 11.6% ISTJ, 8.8% ISFP, 8.7% ESTJ, 8.5% ESFP, 8.1% ENFP, 5.4% ISTP, 4.4% INFP, 4.3% ESTP, 3.3% INTP, 3.2% ENTP, 2.5% ENFJ, 2.1% INTJ, 1.8% ENTJ, and 1.5% INFJ. [7-5]

2018 MYERS-BRIGGS MANDELA EFFECT SURVEY

In 2018, I conducted a survey of 470 individuals, asking people to share their Myers-Briggs personality types, and whether they noticed the Mandela Effect. 418 of those surveyed stated that they had experienced the Mandela Effect. The top Myers-Briggs personality types who reported having had Mandela Effect experiences were: 26.56% INFJ, 25.36% INFP, 9.09% ENFP, 6.94% INTP, 5.5% INTJ, 4.07% ENFJ, 3.83% ENTP, 2.87% ISFP, 2.39% ENTJ, 2.39% ISFJ, 2.39% ISTP, 0.72% ESFP, 0.72% ESTJ, 0.48% ESFJ, 0.48% ESTP, and 0.96% ISTJ with 5.26% stating they fit none of the above categories.

EMPATHS: Intuitive Feelers (NF)

Some people call Myers-Briggs Intuitive Feelers: Empaths, Idealists, or Catalysts. These people comprise the largest group of Mandela Effect and reality shift experiencers. Empaths consist of: ENFJ Teachers, INFP Healers, ENFP Champions, and INFJ Counselors. Idealist personality types adore abstract analysis, and prefer making decisions based on feelings.

As stated before, while Intuitive Feelers comprise only 14% of the general population, they represent a whopping 65% of Mandela Effect experiencers, and are the personality type most likely to

174

report Mandela Effects. This strong correlation between Intuitive-Feelers and the Mandela Effect invites us to investigate and explore this connection further.

RATIONALS: Intuitive Thinkers (NT)

Rationals are Intuitive Thinkers. Rationals include: INTJ Strategists/Masterminds, ENTJ Commanders, INTP Philosophers/Architects, and ENTP Inventors. Notable qualities shared by these types of individuals include: ingenuity, pragmatism, curiosity, innovativeness; they are also logical, calm, strategic, and independent. While priding themselves on being strong-willed, independent thinkers and ingenious problem-solvers, Rationals are chosen by their organizations and groups to be strategic leaders, and are noted for being self-contained pragmatists, who maintain a clear focus on careful analysis of situations and finding optimal, efficient solutions.

Intuitive Thinkers represent about 10% of the general population, yet they represent 18% of Mandela Effect experiencers.

ARTISANS: Perceivers

Artisans include: ESFP Performers, ISFP Composers, ISTP Crafters, and ESTP Promoters. Artisans are noted for being: optimistic, impulsive, daring, enthusiastic, playful, persuasive, adaptable, and charismatic. Unconventional, fun-loving Artisans trust their impulses, placing a high value on freedom, and enjoying being unconventional, bold, spontaneous, and creative. Artisans love variety and new experiences, and being wherever things are happening. Artisans boldly take roads that others might not even consider, unconcerned with standard rules and norms.

Artisans comprise approximately 27% of the general population, and as a group they reported witnessing 6.46% of the Mandela Effect experiences reported.

GUARDIANS: Judges

Guardians consist of: ESFJ Providers, ISFJ Protectors, ISTJ Inspectors, and ESTJ Supervisors. Guardians are noted for qualities of being: dependable, steady, factual, respectable, cautious, logical, and detailed. Guardians pride themselves on being hard-working,

175

helpful, and dependable. Guardians are concerned citizens who trust authority, seek security, and value loyalty. Guardians most respect those with traditional credentials, and are cautious about making changes. Guardians take their duties and responsibilities seriously, and value traditional social customs.

Guardians comprise about 46.4% of the general population, and report experiencing 4.55% of Mandela Effect experiences. Of all four general groupings, these Myers-Briggs personality types are least likely to witness Mandela Effects.

EMPATHY AND PERSPECTIVE SWITCHING

Empathy is the ability to sense other peoples' emotions, imagining what others may be feeling or thinking—and it is a skill that can be developed. People who consider themselves to be natural empaths comprise between 86% to 89% of Mandela Effect experiencers, with the INFJ and ENFJ types noted as being most empathic.

Empaths share an ability to experience consciousness from other perspectives. I am an INFJ type, and I tend to experience reality shifts most vividly when moving my conscious awareness back and forth between awareness of being an eternal spiritual being, and embodied in a physical, mortal body. For those of us who consider the Imaginal Realm to be real, such as shamans, indigenous elders, spiritually faithful, and mystics do, we will tend to experience more frequent and more noticeable reality shifts and Mandela Effects. Moving back and forth between different observational perspectives and levels of conscious agency is something Empaths do naturally, which tends to actuate opportunities for witnessing physically different realities.

The high correlation between Mandela Effect experiencers and the quality of Empathy provides us with powerful clues as to possible quantum mechanisms that we can utilize when seeking to experience reality shifts, Mandela Effects, or miracles. The ability to hold two (or more) different observational perspectives appears to be closely aligned with the ability of Empaths to witness alternate histories.

COINCIDER TRAITS

Over the years, I've noted there exists a strong connection between reality shifts and synchronicity. These often occur in ways that suggest there is some higher hidden order of conscious agency at work. In the "How Do You Shift Reality?" surveys, I discovered that 95% of respondents in 2000 and 93% of respondents in 2013 experienced synchronicity in their daily lives. [7-6]

We might glean additional insights by noticing personality traits correlated with people who experience meaningful coincidences, or "Coinciders," as Bernard Beitman calls them. Beitman found that coincidence sensitivity is related to:

1. **Self-referential thinking** (associating external events as having personal significance),

2. **High emotion** (either positive or negative),

3. A **spiritual/religious outlook**, and

4. **Searching for life meaning**. [7-7]

Self-referential thinking allows for perspective shifting between different levels of conscious agency. This is something that I also found to be key to experiencing Mandela Effects and reality shifts, as well as high emotion, spiritual outlook, and searching for meaning. These qualities can also be seen in the 25 years of first-hand reports in the RealityShifters "your stories" archives.

COMMUNICATING WITH THE APATHETIC

It helps when conversing with people who are disinterested in the Mandela Effect to not bring the subject up directly, but rather follow where their conversational interests naturally go. Over time, thanks to following a laissez faire "hands off" approach, it's more likely that people might take more interest in the subject as it becomes personally relevant in their lives.

Some people can be so skeptical about the Mandela Effect that they adopt an attitude that is dismissive, unkind or sarcastic. Such attitudes indicate that these people are not yet ready to discuss this topic, and may have adopted a reality bubble, or "Wigner Bubble," in which to them, the Mandela Effect does not exist.

While some Mandela Effect-affected individuals may feel tempted to guide others to see the Mandela Effect, such efforts can ultimately feel disappointing. The term "Smith Effect" was mentioned in December 2017 by YouTuber Sovereign Sage, pointing out that people can be pulled in by whatever the current mainstream official version of history might be, rather than investigate their own observations. [7-8] People may recall Mandela Effects to some degree, but not yet feel comfortable acknowledging that histories and facts can change so radically.

INFORMATION FROM THE FUTURE IN THE PAST

From a quantum physics point of view, Mandela Effect experiencers can expect to see some degree of retrocausality with major events, where future choices naturally influence quantum events, since all of reality exists within and can be influenced by "the quantum realm." When we know what to look for, we can recognize the possibility of future influence on many current and past events. We can witness such insights both personally, via our future memories, and sometimes we can collectively observe apparent future messages to our past and present selves.

One possible example includes a novella originally published in 1898 about an "unsinkable" approximately 800-foot cruise liner ship, that hit an iceberg in the North Atlantic on an April night, without enough lifeboats, resulting in the death of most of its passengers. The ship in that story was called "The Titan," and the novella was published about 14 years before the actual *Titanic* sank. [7-9] This weird synchronicity could be viewed as precognition, predictive programming, potential evidence of for instigating future sabotage, or as expanded consciousness connecting via the psychosphere with events in the future and past.

The TV show, *The Simpsons,* doesn't always accurately predict the future, but when it does, it can be breathtakingly brilliant. For example, in the 1998 *Simpsons* episode, "The Wizard of Evergreen Terrace,"

Homer Simpson becomes an inventor, and jots down a rather complex mathematical equation onto a chalkboard. Renowned British science writer, Simon Singh was amazed that Homer's final figure on the chalkboard is fairly close to what physicists discovered about the Higgs boson "God particle" in 2012, stating, *"That equation predicts the mass of the Higgs boson. If you work it out, you get the mass of a Higgs boson that's only a bit larger than the nano-mass of a Higgs boson actually is. It's kind of amazing as Homer makes this prediction 14 years before it was discovered."* [7-10]

In another startlingly accurate prescient moment, a 2000 episode of *The Simpsons,* "Bart to the Future," showed scenes from a fictional future referencing a past presidency of Donald Trump—who didn't actually become president of the United States of America until six years later, in 2016.

SURVIVING & THRIVING IN CHAOS

The Mandela Effect inspires us to embrace an empathic, intuitive, nonlinear perspective and way of living in the world, based on harmonious alignment with higher self, rather than via exerting egoic control. Most initial experiences with reality shifts and Mandela Effects are gentle. Once we trust that we are safe, and are engaged in two-way communication between our thoughts and feelings "in here" and physical reality "out there," we can develop a skill set of staying calm and grounded regardless what seems to be happening.

How each of us adapts with changes to thrive amidst chaos and uncertainty can make a tremendous difference, not just for each of us individually, but also for us in our families, communities, and the world. Our positive attitude and expectations of relaxing in the unknown and trusting Nature and divine Cosmic mind can help to create wonderful new experiences. While each personality type responds a bit differently to challenging times, we can all benefit from honoring each person's unique path.

> **Empaths** who are NF intuitive feelers can easily pick up other peoples energies and emotional and thought-form "stuff." It can be helpful to note that when such discomfort is felt, there may be a good overall purpose, as well as important, timely information. There might, for example, be a call for assistance by someone needing energetic support. During times of turbulence, empaths can notice a palpable sense of chaotic energies—and yet even amidst apparent disarray, notice that connection to a deeper underlying harmonic order can also be felt and strengthened.

Rationals stay centered in trusting their independent thinking, curiosity, and logical strength. Rationals help to ensure that factual analysis, and logical reasoning are the pragmatic, practical foundation of every strategy and plan.

Artisans employ their trusted senses to stay in touch with harmonic flow. Sensory perceptions are trusted friends for artistic types. Amidst times of uncertainty, artists can trust their senses to enthusiastically and daringly surf even the wildest waves.

Guardians rely upon their senses, with the added confidence that judgment brings to help ensure stability, order, and rigorous standard practices. Guardians can also establish helpful practices that are fair, secure, and reliable.

MANDELA EFFECT COMMUNITIES

Initial experiencers of the Mandela Effect often realize that the majority of people in their lives may not be anywhere near as fascinated by the Mandela Effect as they might hope. Thanks to online communities in just about every meeting place on the internet, those who might have difficulty meeting others who experience the Mandela Effect can regularly connect with like-minded people.

From the outside looking in, the Mandela Effect has been branded a *"new and unusually technophilic conspiracy theory."* Mandela Effect experiencers were described by Aaron French as, *"self-identifying Mandela Effect researchers,"* that included New Agers and esotericists, as well as fundamentalist Christians *"unknowingly drawing on the same reservoir of esoteric knowledge."* French acknowledges remembering that the Berenstain Bears used to be spelled "Berenstein," and he felt "vexed" to learn it's now apparently always been spelled "Berenstain," yet he strives to attain an objective outside observer perspective. French described how he researched and confirmed that the official Berenstain family history currently shows that they actually changed their name when coming to America from the Ukraine, in order to sound less Jewish. So while there now exists a historical record of this family name change, it occurred long before any of the popular children's books were written. [7-11]

Aaron French identifies points of similarity and difference between two Mandela Effect conspiracy-focused camps, such that one *"subscribes to a mostly secular conspiratorial worldview filled with corrupt businessmen and politicians trying to enslave humanity,"* while the second camp *"colors their language with biblical references, the dawning apocalypse, ..."* French also

180

acknowledges a third group of Mandela Effect-affected individuals who choose to adopt a more positive viewpoint of the Mandela Effect. French suggests possible directions for future areas of study, in order to explore new memory and hive mind—for the purpose of tracking the Mandela Effect's:

> *"relation to the cultural imaginary of a secularized, technoscientific society. The examples concern the capacity, ability, and limitations of memory, both human and digital, and reveal that memory is more complex and enigmatic than is supposed."*

I would add to French's suggestions for future research, the inclusion of the work of Jerome Busemeyer and Peter Bruza in *Quantum Models of Cognition and Decision*, to incorporate the matter of how human memory operates in quantum ways. We have reached an extraordinary point in time where we see evidence of consciousness behaving with quantum characteristics, so that initial starting conditions and order of questions make a difference, and Nature operates according to quantum logic.

I also suggest further study of languages of cultures such as Irish Gaelic and indigenous groups such as the Hopi, who include in their languages process-oriented quantum approaches to engaging between our subjective internal human experiences and the world "out there." Future study of the Mandela Effect as Extraordinary Human Experiences (EHEs) can also shed further light on the vital role that consciousness plays in our lives.

Awareness of how Artificial General Intelligence (AGI) can begin to become self-aware and exhibit increasing levels of consciousness can also add priceless insights in future directions of Mandela Effect and consciousness research. [7-12]

RELATING TO A CHANGING PAST

> *"May your choices reflect your hopes,*
> *not your fears."*
> —*Nelson Mandela*

Whatever personality characteristics and beliefs Mandela Effect experiencers may uniquely express, all face the essential matter of coming to terms with relating to a changing past.

Many people wonder why the Mandela Effect seems to specialize in the realm of the nearly inconsequential. After all, it hardly matters whether the *Star Wars* robot character, C-3PO, has always been entirely gold, or has a silver leg. It also seems trivial whether the Kit-Kat candy wafer label has a hyphen between "Kit" and "Kat" or not, in the grand scheme of things. The multitudes of Mandela Effect changes being noticed in product logos, spellings of names, and other trivial matters thus seem mostly harmless. When we stop to think about it, we can recognize that these little changes are just about the most gentle way we could become aware that collectively, we might be responsible for them. If it is true that the Mandela Effect provides us with evidence of a Great Awakening, it's wonderful that the first changes we observe are gentle ones.

I've found that maintaining an optimistic attitude, with genuine curiosity to know how good things can get is most helpful in ensuring that personal Mandela Effects are a net positive in my life. The very best results I've experienced have come from being in a positive energetic-emotional state, corresponding to reverence, gratitude, overflowing unconditional love, kindness, and joy.

MANDELA EFFECT MINDSET

As the Mandela Effect reaches a tipping point of global recognition, we can benefit from reviewing the long history of knowledge of intuitive ways of knowing. Only over the past several hundred years has humanity leaned toward favoring a more "left-brain" analytical point of view. The shift of human language from wide and vague to narrow and precise paralleled a movement from metaphorical thinking to a more literal type, as described in beautiful detail in Gary Lachman's *Lost Knowledge of the Imagination.* [7-13] What has been lost in this shift is a sense of respect and trust in our intuition and ability to engage directly in imaginal realms, without being ridiculed for honoring direct intuitive guidance in our lives.

Swedish physicist Dr. Carl Calleman studies overlapping cycles of waves of consciousness, showing that the past four hundred years between 1617 and 2011 corresponded with a wave cycle of "the sixth wave." Around 2011, at the same time that the ninth wave began, we entered the seventh "night" of the sixth wave. [7-14] The significance of this timing might be viewed as showing that the Mandela Effect has arrived to bring awareness to mankind of a newfound mind-matter interactive ability within collective consciousness, and collective history. The perception filters operating in our lives are being revealed to people who are awakening to the Mandela Effect and asking questions, upon noticing that facts and histories are changing. Perhaps the biggest question of all is: Who am I?

THERE IS NO NOISE

A natural phenomenon known as "stochastic resonance" demonstrates that noise can actually aid coherence, which seems a bit counter-intuitive. Typically, when we envision attempting to make sense of what someone is whispering to us across a table, we don't imagine we would hear the message more clearly if we were seated at our table situated outside at a noisy cafe. Yet numerous scientific studies confirm that background noise can be helpful in this way. [7-15] Tony Bell describes how we can see evidence that "there is no noise" in background processes in biological systems, and emphasizes, *"Biological order comes from information flows, up and down levels via emergence and submergence."* Our brains consist of networks of neurons, open to the environment, that communicate among themselves and also through the networks of synapses they are composed of, which in turn communicate through networks of floppy macromolecules. The reason that our bodies can function so well in what at first seems to be a potentially cacophonic din of information exchange is that while there are no levels in biological systems that autonomously compute, there do exist mesoscopic junctions. Bell points out that the main operational principle in our brains and in all living matter is the existence of level switching networks, coordinating exchanges of emergent and submergent information.

A key piece of the Mandela Effect puzzle has to do with the way that some changes we experience contain profound insights, so we can find meaning beyond what initially may seem like noise. We can discover hidden insights in Mandela Effect changes through harnessing intuition, and asking ourselves how a given reality shift or Mandela Effect may show us greater meaning and truth. There can be advantage in maintaining a positive perspective, rather than slipping into doubts or fears, since our emotions can provide a kind of steering mechanism by which we can experience what we are focusing our attention on. We can adopt an optimistic attitude when contemplating the future, and also when considering the past. Changes in past histories that we witness with the Mandela Effect hint that how we focus our attention can also influence what we later discover happened in the past.

Physicist Thomas Hertog pointed out that *"quantum physics forbids a single history,"* inviting us to contemplate implications of the idea that we

183

might never have just one agreed-upon sequence of past events. The extent of the implications of this in our world and personal history are mind-bogglingly huge. Stephen Hawking and Thomas Hertog's idea of holographic quantum cosmology combines elements of both holographic and "many worlds" or multiverse quantum interpretations together, with a smaller range of possible "pocket" universes. [7-16] Taking these ideas seriously means accepting the possibility that there really might be untold numbers of possible pasts for each and every one of us, as well as for the cosmos that we share.

When we view our historical past as being every bit as fluid and malleable as the future, we gain a newfound sense of how we move through time. Once we become aware of the possible existence of multiple histories, we can more readily experience them with all their miraculous potential. And we can also increase our range of experiences with such things as: Déjà Vu, future memory, precognition, premonitions, intuitive hunches, synchronicity, exceptional situational awareness, and spontaneous remissions from injury and disease.

MANDELA EFFECTS IN PERIPHERAL AWARENESS

Some Mandela Effect experiencers wonder why peoples' observations are sometimes different, such that they notice some of the same Mandela Effects, but not all, or at different times. One contributing factor to such differences is the idea of Wigner Bubbles from quantum physics. A Wigner Bubble has to do with the way one observer is observing another observer, with the intriguing possible situation that there may not be agreement by both of these observers on what just transpired—even at the same place and same time. We are now starting to see experimental results confirming that two observational devices at the same place and time can, and indeed sometimes do, observe different events.

There exists a tendency within the Mandela Effect such that some people who regularly observe certain things tend to be unaware of changes in areas they are most familiar with. Subject matter experts are least likely to notice anything has changed within their fields of specialty. It's as if the specific area that an expert is tracking day after day is not making any sudden changes for them, much like, *"a watched pot never boils."*

The Quantum Zeno Effect (QZE) is the quantum mechanism by which quantum watched pots never boil. QZE naturally ensures that whatever state is being continuously observed in a given system will tend to stay locked into that same state. What this implies is that some people notice some Mandela Effects that others don't. People who do not work with human physiology on a daily basis will thus be more likely to remember a time, for example, when there was a reason for the expression "kidney punch," because anyone being struck in the area of the kidneys could be at risk of serious injury or even death.

Some points to keep in mind with regard to Mandela Effects appearing in areas of peripheral awareness are:

1. People will tend to see different Mandela Effects based on areas of interests, with the majority of observed effects tending to be in areas peripheral or just outside of their unique areas of special interests;

2. Mandela Effects are much less likely to be observed in fields of activity in which people are regularly engaged;

3. You can "lock in" particular states of Mandela Effects by more frequently focusing awareness steadily on desired states, and confirming them in those states; and

4. You can increase the number of Mandela Effects you experience by checking the periphery of your focus of awareness more often.

The idea of morphogenic fields that has been popularized by Rupert Sheldrake [7-17], and the type of shared mental ideation described by Walter Freeman, may also have an important role to play in the way we can sometimes more easily access various conceptualizations of reality. [7-18] These ideas illustrate potential mechanisms by which consciousness is operating in ways we are not yet fully measuring.

PHYSICALLY FEELING MANDELA EFFECTS

Some people have shared with me that they feel physical symptoms and sensations before, during, or after experiencing shifts in reality. One person indicated that typically noticeable physical symptoms occurred within 48 hours of experiencing shifts, and added that they noticed that shifts seem to coincide with a couple of things changing at the same time. Dr. Tarrin Lupo noted that some of the most commonly reported types of physical symptoms of Mandela Effect experiencers include: dizziness, light nausea, upset stomach, queasiness, headache, ringing in the ears (tinnitus), dramatic swings in body temperature, disruptions in diet and sleep, and confusion with words, spelling, and grammar. Ringing in the ears was one of the biggest effects reported to Dr. Lupo by Mandela Effect-affected people. Another reported change included a lack of interest in hobbies and activities that people had previously enjoyed. Dr. Lupo notes that some of these symptoms are quite similar to those felt by people experiencing shock—where, for example, people lose their appetite, or feel a sensation like a knot in their stomach.

These types of physical symptoms are not noticed by all Mandela Effect experiencers. I have not noticed most of these physical symptoms aside from ringing in the ears, which might be related to my being a lifelong experiencer.

On the positive side, Dr. Lupo notes that many people report feeling a sensation of energy moving in and out, accompanied by a renewed sense of purpose, and increasingly vivid and more lucid dreams. Some people mention feeling like their cells are vibrating faster, and they can feel them moving around with increasing energy levels and higher vibrational frequencies, as if they are "leveling up." [7-19]

WITNESSING MANDELA EFFECT FLIP-FLOPS

Flip-flops provide us with the opportunity to see something change back and forth from one state to another, each time seeming as if it's always been that way. Witnessing Mandela Effect flip-flops can help us gain confidence in the reality of the Mandela Effect, since they often grab our full attention and involve recent memories where we feel certain we just saw something a different way. Flip-flops can also help us to steer future events by "voting" for the type of reality we'd like to experience, based on our emotional and energetic reactions.

One of the more interesting aspects of flip-flops is that usually the period of time in which flip-flopping is observed is within a few minutes, days or weeks–after which point things seem to settle down. RealityShifters readers have reported there often is a window of time when the switching back and forth occurs, after which a kind of stabilization takes hold. Some flip-flopping can go back and forth between realities a few times, before stability returns, and a particular history is selected.

The following examples of some of my personal Mandela Effect flip-flops are shared here with hopes that they can provide insights and clues as to how these flip-flops can be initially observed, and then interacted with during the flip-flopping superposition of states.

EXAMPLE:
RAIN GUTTER FLIP-FLOPS

In October 2016, I noted that my next door neighbors did not have leaf guard protectors over their rain gutters, but I also remembered that had not always been the case. I looked out my dining room window one day, and saw something surprising that I'd not previously noticed on the adjacent roof–a beautiful leaf-guard system that had been installed atop the rain gutters on my neighbor's house. I did a double-take, while wondering how they'd managed to install such a large system without having made any noise. I figured I must have just been away during the installation time, which surely would have been busy and loud, and left it at that. A day or two later, my gaze naturally wandered along the outlines of my neighbor's roof, and I was stunned to notice that their rain gutter leaf guard system was gone! A while later, I saw the leaf-guards appear again, and then disappear again—and I confirmed with my husband that he'd seen the leaf guard system, too.

EXAMPLE:
DOG CATARACT FLIP-FLOPS

I find it helpful to know "facts" can be in a flip-flopping state when, for example, I voted for our elderly family dog not having cataracts. When I first noticed his eyes looked like they were developing cataracts, I had an instant emotional reaction of *"No!"* while feeling a sense of certainty that his eyes would be just fine–and then the next time I looked at him, his eyes were indeed just fine. A day or two later, my daughter reported to me that there was sad news–our dog was developing cataracts in his old age. I replied that our dog sometimes *looks* like he has cataracts, but then his eyes can be seen to be perfectly clear when we look at his eyes again. I had a similar conversation with my husband a while later. Fortunately, I live in a household where reality shifts are an accepted part of daily life. Once we'd all agreed this was the case, our elderly dog "locked in" the reality with clear vision, and he had no further evidence of cataracts for his remaining golden years. It can feel heartwarming to join together with people sometimes to assist in bringing greater coherence to a favored reality.

EXAMPLE:
DOOR LOCK FLIP-FLOPS

I was camping in August 1998 when I noticed that several times after carefully locking all the car doors, I'd return to the car shortly afterward to find one of the car doors mysteriously unlocked, this was an older 1980s car that did not have electronic door locks. Sometimes it would be the back right door, sometimes the back left door. I felt flustered, because my intention was to keep the car locked, since my purse, money and valuables were inside! After the first time this happened, I made a concerted effort to double-check to confirm that the doors were locked. The first door to appear unlocked was the left, then the right door appeared unlocked, and then the left again! I felt consternated, but also intrigued. Fortunately, when this happened I had only been gone from the car for a short period of time—less than an hour—and the times I was away from the car for more than an hour, the car doors were still locked when I returned. [7-20]

EXAMPLE:
NUTMEG FLIP-FLOPS

While making Gravenstein apple pie for a block party one day, I was sad to see that we were out of nutmeg. I'd hoped to find nutmeg in the cupboard, but I couldn't find it after having checked every container on the shelf. I visualized seeing the nutmeg on the front left-hand side of the cupboard, as I verified visually and by shifting spice containers in that area that it wasn't there. Since I know it's possible to close a cupboard door, do something else, and come back to sometimes find an item that wasn't there before, I did just that. I shut the spice cupboard doors, and walked over to see if I might find the nutmeg on the kitchen counter. It wasn't there, so I returned to the spice cupboard to open the doors and check again, looking where I hoped to find it in the front left section. I then shut the cupboard and repeated the process of busying myself with another inspection of the kitchen counter. I repeated this sequence two or three more times, and then Voila! The nutmeg spice container was on the front left-hand section of the spice shelf, just where I'd originally hoped to find it! Sometimes, I've walked down the hall, and imagined that when I re-entered the room, I was entering a new reality. I find it helps to provide such conscious invitations to literally walk—or jump—into new realities.

REALITY RESIDUE

The term "reality residue" describes lingering "smoking gun" evidence of alternate histories that are no longer part of official historical records. Such anomalies might include, for example, a videotape case or label showing a film title the way some people remember it to be, rather than how such things now officially exist.

Reality residue can help us see confirmation that other realities are out there. This conceptualization about life in a multiverse might give people a sense that "some people have come from certain timelines." Reality residue thus offers us an opportunity to consider the idea that who we are is not so much any physical embodiment of ourselves, as it is our conscious awareness—the aspect of ourselves that cannot be created, destroyed, or measured in completely objective terms.

Usually there's a consistent history in whatever reality we're in, such that most facts that we remember have some form of corroborating evidence that they occurred. The official history might include people like ourselves—Mandela Effect experiencers, reality shifters, who say,

> *"Yes, I've written a description of it, I've drawn a picture of this particular Mandela Effect, I've simulated what I think I saw. It's not the real thing, but it's what I remember."*

We thus most often see and hear people's artistic representations of previous histories. Reality residue usually consists of such things as: misprints, typos, impressions by others riffing off something from memory, and so forth, rather than massive physical evidence that's overwhelmingly persuasive to everyone, including skeptics. This being said, there can be some real doozies of reality residue, such as watching Walt Disney himself, say the magic words, *"Mirror, mirror on the wall"* on "The Walt Disney Christmas Show," which was originally broadcast on December 25, 1951. It's truly amazing to hear and see Walt Disney say these lines the way so many of us remember them from the 1937 animated Disney film, *Snow White and the Seven Dwarfs*.

PLACEBO EFFECT

There is a growing body of evidence to support the way we can harness the power of suggestion to surf between parallel possible worlds. A good deal of mainstream media coverage of the Mandela Effect sweeps Mandela Effects into the dustbin of so-called "false memories" without consideration of the possibility that alternate histories might be real, and that it might share something in common with the truly miraculous nature of the Placebo Effect. While the Mandela Effect is often dismissed in perfunctory fashion, more thoughtful consideration is now being given to news stories about the Placebo Effect. Intriguingly, the "power of suggestion" is credited both with creating "false memories" (which are hardly ever considered to be a good thing) at one end of the spectrum, and with empowering people to enjoy tremendous positive benefits at the other.

Placebo efficacy has baffled scientific researchers by doubling over the past 30 years in the USA for such areas as: spontaneous remission of debilitating health problems, reduced pain, increased stamina, greater cognitive abilities, and much more. Placebos have proven efficacious for surgery, improved scores on tests, improved visual acuity, relief from pain, and improved mobility. Placebos have proven to be effective for horses and dogs, as well as humans. The Placebo Effect is of such tremendous and growing interest that it is now being studied at dedicated research centers, such as the Program in Placebo Studies (PIPS) and the Therapeutic Encounter, hosted at Beth Israel Deaconess Medical Center at Harvard Medical School.

New paradigms take time to be fully adopted. At this dawn of the Quantum Age, we are now witnessing the letting go of old Aristotelian logic, with its three basic principles of: identity, contradiction, and the excluded middle to be replaced by much more comprehensive and complete quantum logic that thrives in uncertainty, possibility, and the unknown.

SUMMARY

While there may not be any perfect stereotypical Mandela Effect experiencer, we can see that such people are more likely than the general populace to share certain attributes, such as being empathic, intuitive, and high-sense perceivers.

While the mainstream media and mainstream science take time adjusting to the new Quantum Age, we can start enjoying some of the benefits now.

* * * * * * *

The next chapter, *Exceptional Consciousness Experiences,* explores exceptional human experiences, toward a goal of better understanding the effect they might have on ourselves and our future.

~~~~~~

EXERCISE:
WHAT MANDELA EFFECTS ARE YOU WITNESSING?

Noting that you're more likely to observe Mandela Effects in the periphery of what you're paying attention to, what kinds of Mandela Effects have you noticed recently? Some areas to consider are changes in: car and product logos, songs and lyrics, geography, human anatomy, and word spellings or meanings.

~~~~~~

EXERCISE:
WHAT IS YOUR MYERS-BRIGGS TYPE?

If you haven't taken the Myers-Briggs personality test, you can go to a website like HumanMetrics and answer questions to get a relatively quick, free assessment. You can also take the official Myers Briggs instrument online at the official Myers-Briggs website. Once you get your results, you can see if your personality type looks right to you. [7-21] [7-22]

~~~~~~

EXERCISE:
EMPATHY TEST

The quality most strongly associated with Mandela Effect experiencers is empathy. You can take an empathy test online at the Greater Good science center at UC Berkeley, and get a general sense of how empathic you currently are, with tips on strengthening your empathy skills. [7-23]

~~~~~~

EXERCISE:
WALK OR JUMP INTO NEW REALITIES

Invite new realities by walking—or jumping—while imagining doors and hallways as representing portals to new worlds of possibility, and longing to know, "How good can it get?"

~~~~~~

"As I walked out the door
toward the gate that would lead to my freedom,
I knew if I didn't leave my bitterness and hatred behind,
I'd still be in prison."

—*Nelson Mandela*

Chapter 8

EXCEPTIONAL CONSCIOUSNESS EXPERIENCES

One of the most thrilling moments of *The Matrix* movie arrives at the end of the film, after Neo exits a phone booth, where he'd said:

"I know you're out there. I can feel you now. I know that you're afraid. You're afraid of us. You're afraid of change. I don't know the future. I didn't come here to tell you how this is going to end. I came here to tell you how it's going to begin. I'm going to hang up this phone and then I'm going to show these people what you don't want them to see. I'm going to show them a world without you, a world without rules and controls, without borders or boundaries, a world where anything is possible. Where we go from there is a choice I leave to you."

Neo then stands outside the phone booth on a busy city sidewalk, looks at crowds of people around him, and then gazes up at the clouds. And then, after a thoughtful few seconds, he flies up like a rocket into the clear sky. In that climactic moment, we get a "blue sky" feeling of how humans might live up to our fullest potential, with an ecstatic rush of freedom and joy. Just as Neo learned to operate outside of habituated patterns and conditioning, we yearn to be unencumbered by worldly concerns, with a glimpse of *imago* (Latin, meaning "image")—the imagined last stage of human evolution.

The Mandela Effect encourages us to develop a sense of trust in what these experiences invite us to know and see beyond habituated beliefs, patterns, and conditioning. Some Mandela Effects appear to be something akin to miracles—when we see previously extinct species return, human anatomy changing for the better, and the idea that anything at all can be instantaneously transformed, such as in first-hand personal healing experiences, that I describe in my book *Reality Shifts*. [8-1] When we observe how some Mandela Effects appear to be assisting human evolution, such as by moving kidneys to a safer location, we may be catching a glimpse of global spontaneous evolution in action, guided by the unseen hand of levels of conscious agency, cosmic order, and divine intent.

Part of this evolution may involve providing us with epiphanies, insights, and downloads that help us discover deeper meaning in the Mandela Effects we observe. We can see some indications that this may be the case from noting key characteristics of those who most frequently witness meaningful coincidences, or synchronicity. Those who most often experience synchronicity have been found by Dr. Bernard Beitman to be characterized by self-referential thinking, high emotion, spiritual/religious outlook, and searching for life meaning.

EXCEPTIONAL HUMAN EXPERIENCE (EHE)

Exceptional human experiences (EHE) are out-of-the-ordinary, anomalous happenings with lasting impact and relevance. When people have especially profound EHE's, individual paradigm shifts of perspective can occur, such that experiencers mark their life story with a "before" and "after" these events.

In 1999, Rhea A. White noted that Exceptional Human Experiences (EHE) were characterized by such qualities as being: initially spontaneous, transcendent, involving a new experience of self, experiences of connection, direct experience of reality that had been taught could not be true, and freedom from any sense of separation. There is often a climactic sensation in EHEs, accompanied by physical sensations including: tingling, swooning, rapid heartbeat, flush, raised hairs, goosebumps, and breathlessness. [8-2]

Rhea A. White identified 240 types of EHEs, which she grouped into seven general categories: death-related, desolation/nadir, encounter, healing, mystical, peak, and psychical. Exceptional human experiences include such things as: Born Aware, ET / UFO experiences, Lucid dream experiences, Mystical experiences, Near-death experiences, and out-of-body experiences.

It is clear to me that Exceptional Human Experiences can also include reality shift and Mandela Effect experiences that shake peoples' views of reality to the core. Mandela Effect experiencers initially spontaneously witness Mandela Effects, which then transforms experiencers' subsequent feelings, attitudes, and thoughts about themselves and others. And this transformation of viewing reality through new eyes is at the crux of every Exceptional Human Experience. A beautiful aspect of EHEs is that they can often provide a greater sense of connectedness with previously undiscovered aspects of self, other people, Nature, and the cosmos.

DREAMS WITHIN DREAMS

Experiencing the Mandela Effect can give us a sense that we may be traveling through realities in a holographic multiverse, via many interconnected possible realities. While we might at first assume that the world is changing around us, and our roles of Actors-Observers are independent from what's happening "out there," we'll benefit from reviewing such beliefs.

The key to mind-matter interaction depends upon the typically under-examined idea of self. No concept could be more crucial for comprehending the true nature of reality, and the science behind the Mandela Effect. In order to truly grasp what is happening when reality shifts or when we initiate some kind of quantum jump, it is necessary to first comprehend what level of self we are operating with. The idea of reality "bubbles" or "Wigner bubbles" suggests that our perspective (mind) influences the physical realities (matter) that we subsequently experience.

Gottfried Wilhelm Leibniz described consciousness as being aware that we are having an experience, at which point we can appreciate having attained a higher order of awareness. This seemingly simple concept is quite profound, since it shows that by noticing, for example, that I am feeling angry right now, I can immediately realize that my true identity is the one who is observing the one who is angry. We are thus capable at any point of taking a step back to a higher level of conscious agency, and in so doing, literally shifting our perspective to a higher order of conscious awareness.

This idea of higher order of awareness is beautifully described in Michael Singer's book, *The Untethered Soul*. Singer shows how we have access in every moment to inner peace, freedom, and well-being regardless of outside circumstances or events. Essentially, there is no good reason to close our hearts to divine love consciousness. Such understanding is possible when we become aware that we are not the voices within our head. None of the voices that we may hear are us— but we each must experience this for ourselves. As Singer says, *"If you're hearing it talk, it's obviously not you."* [8-3]

An extraordinary case of miraculous perspective-shifting is presented by author Anita Moorjani. Moorjani describes her near-death experience of observing many possible realities that she could select between in her book, *Dying to be Me*. At the time she witnessed these possibilities, Moorjani was deathly ill with a case of cancer that her doctors considered terminal. [8-4] Moorjani's healing came as a direct result of her inner experiences.

ANCIENT MANDELA EFFECT MASTERS

Masters of Yoga, martial arts, nei kung, Tibetan Buddhism, and shamanism have demonstrated proficiency in perspective shifting for thousands of years of human history. Yoga masters have extensively documented siddhis, or supernatural abilities, that are bi-products of focused concentration. Yogis demonstrate reality shifts such as: bi-location, teleportation, materialization of objects, invisibility, weather changes, time travel, levitation, and more. [8-5]

Martial artists have been observed in scientific studies to exert mind-matter influence through focused attention and internal energy. Helmut Schmidt's 1976 study and Henry Stapp and Helmut Schmidt's 1993 experimental studies showed that martial arts subjects successfully influenced previously recorded rates of radioactive decay—effectively changing events in the past. [8-6] [8-7]

Nei kung masters such as John Chang, also known as "Dynamo Jack," have been observed doing amazing things, such as: healing with his hands, catching a pellet from an air rifle in his hand, pushing a chopstick through a tabletop, and starting fires with energy from his hands. [8-8]

In 2004, I attended a talk by Mongolian shaman Zorigtbaatar Banzar, about how shamans in his culture affect healing, change weather, shapeshift, and more. Zorigtbaatar stayed up late at night, talking with several of us about how to best ensure optimal transformation of human consciousness. He explained that we can ensure best results by forming cohesive groups to care for and nurture one another, without becoming disconnected. Zorigtbaatar pointed out that we are all directly in touch with the Absolute, with shamans actively directing Galactic, Earth, and human energies in positive, healthy ways. Shamans do this by focusing attention on underlying energy dynamics. [8-9]

DOZENS OF WAYS TO SWITCH OFF LEFT BRAIN

While Exceptional Human Experiences including the Mandela Effect are typically initially spontaneous and unplanned, we can enter into a more receptive state of mind to encourage Mandela Effects in our lives. One helpful method involves finding ways to quiet the ceaseless chatter of our left brain.

Many of our world's top scientists, musicians, and artists engage the creative power of right-brain thinking when receiving sudden flashes of insight. Just as Paul McCartney received his song, "Yesterday" in a dream, some of the top scientific discoveries such as seeing the double helix structure of DNA have also come through in the form of direct knowing and inspiration. Canadian author and award-winning UFO researcher, Grant Cameron, shows how we can gain such insights by utilizing dozens of contact modalities, such as meditating, that provide us with many ways we can choose from that can help to quiet the left side of our brains and enhance intuition and creative insights. Cameron provides over 70 methods by which people can transcend ordinary ways of rational, analytical knowing, in order to benefit regularly from switching off our hyperactive left-brain thinking, and gaining many benefits from directly connecting with consciousness. [8-10] It seems clear to me that reality shifts and the Mandela Effect are related to what Cameron describes as "Apports" in his Alien modalities section. Apports need not be directly related to UFO or ET contact experiences, and many people who witness items appearing out of "thin air" also experience those items vanishing again, or transporting, or transforming. Clearly, the phenomena of reality shifts and the Mandela Effect qualify as contact modalities, with the recognition that these experiences can be fostered by switching "left-brain thinking" off.

It is possible that our experiences with the Mandela Effect may grow and change along with us. Cameron describes how UFO experiences have changed over the centuries, from wooden ships seen in the air in the 1800s to modern-day experiences with orbs, mists, beams of light, and triangles.

Humanity now has an opportunity to experience what a difference our beliefs can make, when we consider that the foundation of reality may be consciousness—and not physical matter—and our personal experiences with the Mandela Effect can be a beautiful invitation to make this journey. We can enjoy playfully engaging with reality. We can reclaim gifts that our ancestors knew about, and built into human languages. We can participate in creation at a higher level of conscious agency, where mind-matter interaction becomes increasingly seamless.

PLAYFUL CREATIONS IN SHIFTING REALITIES

Many types of playful creations are possible with personal Mandela Effects, reality shifts, or quantum jumps. Over the past twenty-five years, dozens of people have shared with me their experiences of traveling farther in less time than should be possible, without speeding or doing anything unusual. They arrive at destinations hundreds of miles away in half the driving time required, wondering how such a thing could be possible. This type of time shift is one of the most commonly reported reality shifts, and it's a fun one. Some experiencers make a point to state, *"I was not speeding at any time,"* while also noticing that they would have had to be breaking land-speed records in some cases, in order to arrive at their destinations when they did. Other types of time shifts include such things as: time slips, déjà vu, repeating cycles of events, and future memories.

Another commonly enjoyed type of personal Mandela Effect involves items reappearing that had previously inexplicably vanished. Many such reports have been documented in the realityshifters archives. These JOTT or *Just One of Those Things* incidences have been commonly experienced for centuries, as reported by Mary Rose Barrington in her book by the same name, and in publications for the Society for Psychical Research. [8-11]

We can experience heightened levels of personal Mandela Effects when feeling intense emotions, such as love, together with awareness of different perspectives. I experienced an emotionally tumultuous month of transition in August 2008, when I was helping my older daughter prepare to move away to college. I was surprised by feeling pangs of loss long before she left, with an intensification of heartache the closer we came to the day she would move away. What I found most surprising was feeling emotions in my entire body—such as heartache, butterflies in my stomach, and nervous restlessness. It felt like my love for my daughter was so vast it could not fit in my heart; it affected all of me. Before going on my morning walk near the woods in the early morning summer fog in August 2008, I got a clear intuitive future memory insight that I would see things I would like to photograph, and it would be a good idea to bring my camera with me. With camera slung over my shoulder, and the family dog trotting happily at my side, I wondered what I might find. My thoughts turned to those I love, and how elusive the most beautiful things can be to capture on film, due to their fleeting, transitory nature and beautiful essence that can seldom be collapsed into two visual dimensions. With my mind wandering, I completely forgot about taking pictures, as I admired nature's beauty all around me.

I walked along the forest and looked up from my reverie, stunned to notice a heart-shaped stump. I had walked past it dozens of times before, yet today was the first time I had noticed the heart shape—so I stopped and took a picture. Continuing my walk, I found dozens more hearts in the most unlikely places, and snapped photos of heart-shaped ivy leaves, heart-shaped lichen growing on a rock, a heart-shaped string arranged perfectly in the road, a heart-shaped stone, and heart-shaped branches in an oak tree. Everywhere I looked, there was another heart looking back at me. Some were less obvious than others, such as a little heart-shaped-string resting on the asphalt, hinting at the essential form of a heart, as if to say, *"You know what this is!"* and *"Wherever you go, there is love."* [8-12] With each joyful discovery, I became aware of how focusing on love can bring showers of hearts in great profusion, providing confirmation that simply by feeling so vulnerable and open to experiencing the vastness of love in my life, here it is, all around. All we need do is recognize it when it shows up in its many forms and guises.

Many people have shared stories with me for RealityShifters about how they find profusions of lucky pennies and four-leaf clovers. One of my best friends finds heart-shaped rocks everywhere she goes, to the point that her father once commented, *"Are all rocks heart-shaped?"* These discoveries represent similar examples of finding things when in the right state of mind. Such reminders give us a sense of connectedness and love, when we might otherwise feel alone. When something shows up as a sign, again and again, we can recognize we are living lucidly in ever-shifting realities that are teeming with symbolic insights and messages.

QUANTUM RETROCAUSALITY IN IRISH GAELIC

The Irish Gaelic language may seem archaic to some, but it is actually experiencing something of a modern day revival, as its surging popularity on online training programs attests. Irish Gaelic includes a special "imprecatory" mode, that corresponds beautifully to the idea of delayed choice that physicist John Archibald Wheeler considered possible within quantum mechanics. The idea of delayed choice involves changing what occurred in the past, and an example of such a statement in Irish Gaelic is:

"Dá dtogadh sé 5 in ionad 1!"

which is an after-the-fact prayer for events in the past, that someone did not take Highway 1 between Los Angeles and San Francisco (which we're imagining we'd just heard has fallen into the Pacific ocean), but instead took Highway 5. This statement is made in Gaelic in the imprecatory past—holding a positive vision for an event that has already occurred. [8-13]

Language has a subtle yet profound influence on our beliefs and worldview. When people grow up speaking Gaelic, a philosophical benefit that comes along with the language is a sense of knowing that we can hold good thoughts in mind for all uncertain events, in the present, future, and past. It helps to not yet know what has happened, in order to do the best job of holding a positive vision for past events.

HOPI ENTANGLEMENT AND COHERENCE

The Hopituh Shi-nu-mu "the peaceful little people" of northeastern Arizona, have beautiful words and a beautiful way of thinking to describe how change is introduced into our lives. *Tunatyava* means, "comes true, being hoped for," an expression of gratitude and confidence in the way the Great Spirit brings to us what we feel in our hearts. This humble acceptance of what is and what has been honors the interconnectedness of all things without questioning or measuring the time required for a change to occur.

The Hopi concept of *tunatyava* represents fundamental comprehension of the way probabilities become realities. While the English language places events at specific points in time, the Hopi language operates outside of linear time references, and has no words or expressions to refer to time. The Hopi language has no verb tenses to identify past, present, or future. The Hopi consider the world to consist of either that which we experience as manifest (*tunatyava*) or that which is in our hearts and which we are actively manifesting (*tunatya*). Such a conceptualization is remarkably similar to physicist David Bohm's ideas of implicate and explicate order in a holographic cosmos. The manifested tunatyava includes all that is and has ever been accessible to the senses. The manifesting *tunatya* incorporates the moment of inception in the present and all that we consider future, all we consider mental, and everything that exists in the hearts of animals, plants, people, things, nature, and the cosmos itself.

The idea of *Tunatya* is thus a dynamic state of infinite, motionless creativity connecting through us to bring into being what is most desired. *Tunatya* is the required state of mind and consciousness by which the Hopi bring rain to crops that have no irrigation system, based on living in a state of respect for Spirit, plants, animals, the land, and one another. There is awareness of the interconnectedness among all of these, with clarity of knowing we cannot separate the effects of our thoughts, words, feelings, and actions—that they must be integrated and based in respect and reverence for best results. The

Hopi believe humans have moved through four worlds so far, surviving times of great global upheaval, including a great flood. According to Hopi mythology, we now reside in this World Complete, or *Tuwaqachi*. We will see the signs of the emergence to the fifth world when life forms from previous worlds spring up from the ground, the stars, and our hearts. [8-14]

Viewing Mandela Effects while considering the Hopi idea of the Fifth World, we see that many indigenous peoples have been awaiting and expecting the arrival of new and changed animals. The Hopi expect the return of extinct and endangered plants and animals to be one of the signs indicating the threshold of leaving the Fourth World and entering the Fifth World. The Hopi advise people to be mindful at times of moving from one world to the next, which can be times of great change and uncertainty. The key to success lies in heart-centered care for relationships at times of shifting from one world to the next. The most important thing is to stay connected with Creator while simultaneously keeping our hearts open with love, kindness and compassion. These ideas are illustrated on the Hopi Prophecy Rock, indicating one dead end path, based primarily on extrinsic (outer) rewards—and one path going forward that does not end, honoring heartfelt relationships to Nature and one another. [8-15]

LAZARUS SPECIES

Lazarus is a name given to species that have been rediscovered after having been presumed to have gone extinct—sometimes for a decade, and sometimes for millions of years. Like Lazarus of Bethany, some animals thought to have been gone forever from this Earth can sometimes be comeback kids. Another possibility to consider when an entire species dies out and later is "Alive Again," sometimes after tens of thousands of years is that we're experiencing the Mandela Effect.

A partial list of some species previously considered to have gone extinct that are confirmed as having returned include: Horned Marsupial Frog (Gastrotheca cornuta), Fernandina Giant Tortoise (Chelonoidis phantasticus), Coelacanth (Coelacanthiformes), Black-footed Ferret (Mustela migripes), Bermuda petrel (Pterodroma cahow), Chacoan peccary (Catagonus wagneri), Lord Howe Island stick insect (Dryococelus australis), Monito del Monte (Dromiciops australis), La Palma giant lizard (Gallotia auaritae), Takahe (Porphyrio hochstetteri), Cuban solenodon (Solenodon cubanus), New Caledonian crested gecko (Rhacodactylus Ciliatus), New Holland mouse (Pseudomys novaehollandiae), Giant Palouse earthworm (Driloleirus americanus),

Large-billed reed-warbler (Acrocephalus orinus), and the Laotian rock rat (Laonastes aenigmamus). [8-16] [8-17]

The horned marsupial frog, which incubates eggs on its back, disappeared for more than a decade before it was rediscovered. [8-18]

The Fernandina giant tortoise had last been seen in 1906, and had been thought extinct—before scientists were stunned and thrilled to find one in the Galápagos in 2019 on Fernandina Island. This female specimen of *Chelonoidis phantasticus,* "the fantastic giant tortoise," was found on an isolated patch of vegetation on the southeast of the island, cut off by several lava flows. Recent signs of tracks and scat of at least two or three other tortoises have been found in expeditions on the island. [8-19] [8-20]

The coelacanth (pronounced SEEL-a-canth) is possibly one of the most famous Lazarus creatures. Measuring the size of a large male human, coelacanth fish can weigh about 200 pounds, and measure 6.5 feet long when fully grown. With lifespans of 60 years and living 2,300 feet below the waves, these primitive-looking fish had been thought to have gone extinct 65 million years ago with the dinosaurs. Its discovery in 1938 in South Africa, and later in Indonesia, thrilled and amazed people around the world. [8-21]

Another animal thought to have gone extinct 11 million years ago is the Laotian rock rat. It was rediscovered in 1996, when scientists considered it to be so unusual and different from any other living rodent that at first it was given its own family. Ten years later in 2006, the Laotian rock rat was reclassified to belong to an ancient fossil family. [8-22]

In November 2018, research associate Jeff Goddard was seeking nudibranch sea slugs at Naples Point, California, when a pair of small, translucent bivalves caught his eye. These tiny creatures measured just 10 millimeters long, but the moment they extended and began waving a bright white-striped foot that was longer than their shells, Goddard quickly took photos of the little clams, which appeared to be rare. It took Goddard many more trips to find a specimen to analyze, and finally, in March 2019, he succeeded. Goddard was amazed to find a new species, which he presumed it to be, since he was extremely familiar with all species along the southern California coast, which he'd been researching since 2002. He was thus stunned to learn this clam was the first living example of the *Cymatioa cooki*, which had presumed to have gone extinct some 40,000 years ago. *"It's not all that common to find alive a species first known from the fossil record, especially in a region as well-studied as Southern California,"* Goddard said. *"Ours doesn't go back anywhere near as far as the famous Coelacanth or the deep-water mollusk Neopilina galathea —representing an entire class of animals thought to have disappeared 400 million years ago—but it does go back to the time of all those wondrous animals captured by the La Brea Tar Pits."* [8-23]

Some animal species appear seemingly out of nowhere, even surprising the experts. Biologist Clint Laidlaw was bewildered when he first saw Beaked Whales, since he was an expert in the field of marine mammals in the order, *Cetacea*, yet Beaked Whales were completely new to him. Laidlaw said, *"I was very surprised a few years back, when I discovered this: a 13 meter, 15 ton whale, with the face of a dolphin. What the heck was this thing, and how could I have gone my whole life as an animal fanatic with multiple degrees in biology and zoology, without noticing?"* [8-24]

Perhaps one of the most remarkable developments of Lazarus species is the way that some biologists now acknowledge that, "sometimes, extinct species return" when habitats and eco-systems are restored. This is a remarkable admission that sounds very much like, *"If you build it, they will come,"* to quote a line of dialogue from the movie *Field of Dreams*, that many of us remember.

ORIGINAL THINKING

How would you answer the question, *"Is it possible to come up with an original thought?"* According to indigenous ways of thinking, the answer to this question is "no," since all possible answers exist within Nature. Within traditional indigenous ways of thinking, it is not our egoic voice or actions motivating us, but something far deeper, from the spirit or soul. From such a soulful place, we can regain a sense of true inner harmony, and genuine interconnection with all of the cosmos.

Glenn Aparicio Parry's book, *Original Thinking*, explores the concept of living with a deep sense of relationship in time and space. To indigenous peoples, time is not linear, and separation is an illusion. Every living thing relates to others, such that events are intrinsically interconnected through space and time. [8-25]

We can see from various interpretations of quantum physics that all possible realities can be viewed as already existing. This corresponds beautifully with the ideas held by indigenous cultures who appreciate that rational thought is a subset of a greater awareness, and restored balance is possible when we allow for larger awareness to move through us. When we have such an awareness, our days and relationships feel rich with meaning and purpose.

UNITY CONSCIOUSNESS

Many people who've had exceptional human experiences are aware of unity consciousness—a state of blissful being-ness with the Absolute. Unity consciousness feels to me like being in Heaven, basking in infinite divine unconditional love. The idea of Unity consciousness is not associated with any particular religion, yet it can be viewed by those of a spiritual nature as a way of conceptualizing Divine Source, the Holy Spirit, Creator, and God. Unity consciousness provides a way of thinking and talking about the underlying hidden order we can learn to feel and recognize in our daily lives.

On the way to directly experiencing Unity consciousness, we may notice an expanded sense of consciousness, of being much more than just a physical body and egoic self. From this sensation, we might glean insights as to how we don't show the world the fullness of all we are. We present a facet of ourselves, often without remembering all aspects of who we've been, or are capable of being.

OUR MULTITUDE OF SELVES

We can envision ourselves as energy beings—as beings of consciousness far greater than could ever fit inside any given physical body. We can gain a sense of the vastness of ourselves through dreams and daydreams, when accessing such imaginal realms.

In any gathering of two people A and B, we will at minimum have not two but four definitive personas present: the person A believes A to be, the person A believes B to be, the person B believes A to be, and the person B believes B to be. In truth, there are far more likely a great deal more personas present even than that, as a fascinating novel by Italian novelist Pirandello vividly describes. The masterpiece that is Luigi Pirandello's *One, None, and a Hundred Thousand* is at first glance the story of a man viewed by his friends and family to have gone mad, as he systematically sets out to destroy every one of his identifiable personas. The genius of this book is how it brings us inside the mind of it's main character, an Italian man named Vitangelo, who becomes obsessed with the notion that his version(s) and understanding of himself are not at all the same as the personas that others believe him to be—nor are they likely to be who he thinks them to be. [8-26]

James Fadiman and Jordan Gruber provide evidence of our true identity as consisting of multiple selves, rather than the presumption of a single self, in their book, *Your Symphony of Selves*. We intrinsically know that we think, speak, and behave differently in different circumstances. We can thus prepare ourselves most effectively to deal with various situations by shifting in advance to an efficacious, optimal version of ourselves, rather than feeling we've had "our buttons pushed" or ending up reacting to emotional triggers that are less-than-ideal at that juncture. [8-27]

We can glean fascinating insights into how we experience our identity from scientific case studies with people diagnosed with Dissociative Identity Disorder, previously known as Multiple Personality Disorder. A blind female patient in Germany spontaneously regained eyesight in all but two of her ten personalities, baffling her doctors. Doctors utilized a tool to measure electrical activity in the woman's brain in response to visual stimuli, and observed that the visual areas of her brain remained inactive when she was functioning in her blind personality, and were active in her sighted personalities. Purely physical models of reality offer little in the way of explaining how such a physiological condition could occur. [8-28]

Such striking changes are evident that as a person shifts from one personality to another, it can almost appear the person is donning a new face. Other studies involving people diagnosed with Dissociative Identity Disorder have involved similarly striking physiological differences for a person in one physical body who might have a personality who: was or wasn't allergic to a particular food, responded completely differently to a given medication, had blood pressure of 150/110 or 90/60, had one eye rotated outward to one side or forward, or had a different eye color. One of the doctors involved in studying these cases notes that,

"A given state of consciousness has its biological reality. By keeping these states separate and distinct, the patients create biologically separate selves." [8-29]

Each of us can be considered to be a multitude, on a scale that can be mystifying to contemplate. Reality does not feel like a simulation in the sense of people being non-player character NPC's and having meaningless relationships with me, but rather precisely the opposite. I feel everything has a degree of sentience seldom fully appreciated in our modern world. I can sense my higher levels of conscious agency, or conglomeration of possible selves, which seems a lot bigger than could ever fit into a singular egoic identity. When I feel a change in perspective between egoic and higher self—and allow my consciousness to move back and forth a few times—I can greatly increase the number of reality shifts and Mandela Effects I experience.

MEETING OUR FUTURE SELVES

Amidst thousands of first-hand RealityShifters reports received since the 1990s, numerous accounts of people meeting themselves have been documented. Historically, two cases stand out as being especially noteworthy. German author and statesman Johann Wolfgang von Goethe described once meeting his future self around 1770, in his autobiography, *Poetry and Truth*. Goethe had just escaped from, *"the excitement of a farewell,"* while noting he was feeling, *"very uneasy."*

> *"Yet amid all this hurry and confusion I could not resist seeing Frederica once more. Those were painful days, whose memory has not remained with me. When I held out my hand to her from my horse, the tears stood in her eyes, and my heart was heavy. I rode along the footpath toward Drusenheim, and here one of the most singular forebodings took possession of me. I saw, not with the eyes of the body, but with those of the mind, my own figure coming toward me, on horseback, and on the same road, attired in a dress which I had never worn. It was pike-gray, with somewhat of gold. As soon as I shook myself out of this dream, the figure had entirely disappeared. It is strange, however, that eight years afterward I found myself on the very road, to pay one more visit to Frederica, in the dress of which I had dreamed, and which I wore, not from choice, but by accident. Whatever one may think on such matters in general, in this instance my strange illusion helped to calm me in this farewell hour."* [8-30]

American author Philip K. Dick described an experience in which he once met his future self—and his wife saw his future self, too. The experience frightened PKD's wife, yet PKD was remarkably calm about this event that transpired around 1951 or 1952:

> *"Back at the time I was starting to write science fiction, I was asleep one night and I woke up and there was a figure standing at the edge of the bed, looking down at me. I grunted in amazement and all of a sudden, my wife woke up and*

started screaming, because she could see it, too. She started screaming, but I recognized it and I started reassuring her, saying that it was me that was there and not to be afraid. Within the last two years—let's say that was in 1951— I've dreamed almost every night that I was back in that house. And I have a strong feeling that back then, in 1951 or 1952 that I saw my future self, who had somehow, in some way we don't understand—I wouldn't call it occult— passed backward during one of my dreams now of that house, going back there and seeing myself again." [8-31]

As described in Chapter Three, when I met my future self, I was stunned to see her float through my mirrored closet doors, glide across the room to my desk, reach down into the lower left drawer, retrieve something—soothingly communicating telepathically with me the entire time—and return the way she'd come. Over the years, I've recognized a possible motive by which a future self might have felt a need to take the first letters that my boyfriend (and future husband) gave to me. The way I see it now, that theft (my future self taking my teenage letters) gave me permission to recognize at some future point that this relationship doesn't have to be forever—that it was OK to get divorced. So my younger self received a message from my future self that would have profound meaning and significance for me for years to come. The fact that I've not yet received any special delivery of those old letters—despite owning the same roll-top desk for decades, and occasionally checking to see if they're there—lends credence to the concept of closed time-like curves, such that histories do not necessarily need to converge into one consistent timeline.

MANDELA EFFECT MESSAGES AND MEANING

We can glean insights from Mandela Effects by analyzing them with similar techniques and methods that we utilize for studying and finding meaning in dreams. We can follow a four step process for dream interpretation that also works for interpreting Mandela Effects that we feel especially intrigued by, as mentioned in Chapter Two.

(1) **Write a narrative** of your experience. Write down the sequence of events that includes some kind of reality shift or Mandela Effect, the way you would describe it to a friend who wasn't with you at the time.

(2) On a fresh piece of paper, **write each symbolic element** of your experience, one word at a time (every significant noun, verb, and adjective) on the far left side of the page, leaving lots

of space to the right of each word, and skipping lines to leave lots of extra space.

(3) Go back to the top of the list of words from step two, and utilize a process of **free-association brainstorming** to document everything that comes to your mind in response to each key word, generating many words and phrases to correspond with each of the dream symbol key words from step two.

(4) Ignoring the narrative from step one, and the key words from step two, read through all the words and phrases from step three, and **create a new summary narrative**. This is your Mandela Effect symbol interpretation, sharing insights and meaning from the experience just for you.

This approach to revealing hidden meaning involves opening dialogue with deeper, subconscious awareness of what is really going on. Key ideas are like dream symbols, and free association can reveal insights you might not otherwise see. Exceptional consciousness experiences—like the Mandela Effect—invite us to more fully participate in enjoying creations of consciousness in our daily lives.

* * * * * * *

The next chapter, *Mandela Effect Hero's Journey and Gifts,* illustrates the way that the Mandela Effect-affected can be viewed as going on a hero's journey, where the process of discovery is the greatest treasure of all. We will gain insights into how taking a hero's journey can make a difference for the hero, the immediate community, and the entire world.

~~~~~~~~~~~~~~~~~~~~~~~~~~~~~~~~~~~~~~

EXERCISE:
INVITE GUIDANCE BY YOUR FUTURE SELF

Before meditating, napping, or going to bed at night, you can invite guidance by your Best Possible Future Self. Imagine that in your meditation or sleep, you can become aware of untapped possibilities in your fullness of possible selves, which are far more than you typically envision.

~~~~~~~~~~~~~~~~~~~~~~~~~~~~~~~~~~~~~~

EXERCISE:
PLAYFUL CREATIONS

Think of some things you'd be surprised and delighted to find—whether a kind of coin, or heart-shaped objects or rocks. Notice how you can find these any and every where!

~~~~~~~~~~~~~~~~~~~~~~~~~~~~~~~~~~~~~~

EXERCISE:
EXPERIENCE UNITY CONSCIOUSNESS

Imagine meeting a vast, unified Absolute consciousness. However this essence of Unity appears to you, know that you can recognize and receive the aspects that you most need right now.

~~~~~~~~~~~~~~~~~~~~~~~~~~~~~~~~~~~~~~

EXERCISE:
ANALYZE A MANDELA EFFECT

Think of a particular Mandela Effect that feels emotionally significant to you. Follow the four-step Mandela Effect interpretation process: (1) Write a narrative of how you experienced this Mandela Effect. (2) On a fresh piece of paper, write down the key ideas (nouns & verbs) from the narrative in step one. (3) Review the key ideas from step two and do free-association, writing down whatever first comes to mind for each of these key words.
(4) Reading through the results from step three, rewrite the narrative of what happened, this time with a sense of what meaning this Mandela Effect has for you.

~~~~~~~~~~~~~~~~~~~~~~~~~~~~~~~~~~~~~~

"We can change the world
and make it a better place.
It is in your hands
to make a difference."

—*Nelson Mandela*

Chapter 9

MANDELA EFFECT HERO'S JOURNEY & GIFTS

The Mandela Effect can be experienced as a call to adventure, inviting us on a journey where we meet mentors, cross a threshold into the unknown, find helpers, gain new skills, and embrace new gifts and revelations when returning home, transformed. American author Joseph Campbell encouraged us to view our lives as heroic adventures, discovering what excites us, while honoring a sense of passion central to our personal journeys. Even as accidental adventurers, we can benefit from revising pre-existing conceptions and limited world visions. New ways of thinking about oneself in relationship to others and the cosmos are an integral part of the hero's journey. Campbell points out,

> *"There's a big transformation of consciousness that's concerned. And what all the myths have to deal with is transformation of consciousness. That you're thinking in this way, and you now have to think in that way."* [9-1]

Many Mandela Effect experiencers undergo challenges in personal relationships, since friends, family, and coworkers may not be interested in exploring Mandela Effects. Viewing the Mandela Effect as a transformative journey can be greatly reassuring. We can find gifts and discoveries amidst the mystery that is the Mandela Effect, by looking at each phase of the Mandela Effect Hero's Journey.

CALL TO ADVENTURE

The first part of the hero's journey is the call to adventure, which is often initially rejected or refused. We often feel safe and cozy right where we are, and not eager to move outside of our comfort zone. Because some reality shifts and Mandela Effects can resonate so deeply and profoundly that they cannot be ignored, they can initiate an awakening experience that nudges us out of our accustomed way of thinking and living.

An example of a call to adventure appears in the movie, *Star Wars*, when Luke Skywalker chooses a robot, R2-D2, for his family. That simple shopping trip initiates a series of unforeseen cascading events that ultimately challenge Luke to step up into experiencing a new world, and a whole new version of himself.

Sometimes the call to adventure can feel scary. There can be a sense of feeling completely alone. We might notice that the people around us feel different, through their disinterest in what we have begun to see is going on.

You'll know that you have received a call to adventure if you feel an unmistakable sense of excitement, or a feeling that from this moment on, nothing in your life will be quite the same. You may feel a call to adventure when witnessing a Mandela Effect flip-flop, when experiencing a Mandela Effect with other people, or when experiencing a shift in reality you cannot dismiss. The hero's journey we get is the one we are ready for. Sometimes trials arise that seem overwhelming, yet we learn to find courage when facing our fears.

MEETING MENTORS

Mentors make a big difference once the hero's journey begins, as we leave our comfort zone. Some mentors on my journey of exploring mind-matter interaction include authors and consciousness researchers: Dan Moonhawk Alford, PMH Atwater, Cleve Backster, Larry Dossey, Stanley Krippner, Bruce Lipton, Leroy Little Bear, Hamish Miller, Jeffrey Mishlove, Edgar Mitchell, Jacob Needleman, Dean Radin, Rupert Sheldrake, Henry Stapp, Russell Targ, George Weissmann, and Fred Alan Wolf. Their influence continues to help me clarify and communicate my observations and ideas.

Mentors can include anyone in your life who you intuitively trust and respect. The best mentors are: compassionate, available, patient, positive, and kind—providing you with a sense clarity and inspiration. Mentors can provide wisdom that helps when new challenges are faced, and encourage expansion and development of our best qualities.

While it helps to work with a mentor you know personally, it's also possible to receive mentorship from people whose work you only experience via videos, social media chat rooms, email, and books. Mentors for Mandela Effect-affected individuals can include authors of books about the Mandela Effect, YouTube creators, and speakers at Mandela Effect conferences and events.

Star Wars film director George Lucas consulted with Joseph Campbell when writing the story for his new movie, *Star Wars*, to master the art of telling stories that engage the audience, through feeling authentic and engaging. Lucas told Moyers,

"I think my last mentor probably was Joe Campbell, who asked a lot of the interesting questions and exposed me to a lot of things that made me very interested in a lot more of the cosmic questions and the mystery." [9-2]

CROSSING A THRESHOLD INTO THE UNKNOWN

You can tell you are crossing a threshold into the unknown when you've arrived in unfamiliar territory. You know you're on the edge of the unknown when experiences feel so foreign that your worldview is shaken to the core. This threshold crossing moment is felt in the first *Star Wars* movie (*episode IV: A New Hope*), when Luke and Obi Wan enter the Cantina, filled with alien beings. Luke and Obi Wan are seeking a star cruiser and pilot, and Luke realizes that his life is about to make a radical departure from his past, as plans for imminent departure from his home planet are made.

Walking in two worlds of experiencing a deeper level of reality while simultaneously living an ordinary life can feel unsettling and unforgettable, as a threshold is reached that cannot be denied. My Mandela Effect threshold moment happened the day I was with two friends and we all witnessed the appearance of a giant concrete sundial sculpture that had never been there before near the Berkeley Marina. Once I had confirmed with friends that reality shifts were real, I could no longer pretend otherwise.

Mandela Effect experiencers reach a threshold of experiencing the undeniable truth that we are living in an extraordinary world. What seems reliable and solid can—and often does—change. Learning to navigate through such ever-changing landscapes can bring us back to our spiritual foundation, where we can find deeper faith in sensing the presence of infinite possibility beyond the illusory confinement of linear, material spacetime.

TRIALS, TRIBULATIONS, & HELPERS

Helpers can assist the hero in getting over the egoic self and gaining a sense of becoming a better person. Growth comes in times of great challenges, where we face choices we might not otherwise have considered. Growth and challenges both show us that we can become preoccupied with egoic matters, or we can shift our awareness to become conscious of a more expansive, higher order perception of ourselves, where we can be of greater service. Adversity provides us with catalytic opportunities to break free from habitual egoic patterns of the past.

Helpers on the journey allow us to broaden our perspectives in ways that are essential for thriving. When we connect with other Mandela Effect experiencers, we can learn from their experiences, ideas, and points of view.

Through interactions with others, we can discover the value of becoming the best possible versions of ourselves. We can learn to appreciate and forgive ourselves and others, and choose to trust, respect, and be kind and compassionate, as we become a person of noble integrity and courageous vision. Heroes seldom start out on the journey in their highest form, but rather find their best selves on the journey. Heroes accept invitations to face emotional and energetic shadow issues and plunge through depths of dark nights of the soul into the dawn of unconditional divine love. Heroes attain new heights thanks to assistance from helpful people in their lives.

The movie, *Sliding Doors,* shows an example of how a woman, played by Gwyneth Paltrow, experiences different parallel worlds of possibility when she makes choices in how she chooses to interact with key people in her life with very different outcomes in two possible parallel versions of her life. Another wonderful movie, *Groundhog Day,* provides dozens of side-by-side alternate realities that the character played by Bill Murray experiences while caught in a seemingly never-ending repeat of reliving the exact same day. In both of these films, it's possible to see beautiful gifts granted to the protagonist when they work toward personal development with the assistance of those around them, finding ways to be in the highest degree of service that they can provide.

GROWTH & NEW SKILLS

In the movie, *Star Wars*, Luke is guided by his mentor, Obi Wan Kenobi, to *"let go your conscious self and act on instinct."* This advice returns later in the movie, when Luke remembers his mentor's advice to, *"Use the force, Luke. Let go."* This advice is a call to allow intuitive guidance to provide direction.

This scene in the movie resonated deeply with me when I first saw it, and I now appreciate how I regained connection with extensive intuitive abilities after my intense kundalini awakening experience in Autumn 1994. During that time, I awakened at 1:11, 2:22, 3:33, 4:44 and 5:55 every night, seeing those repeating numbers on my digital clock. About six months before this happened, around the Spring Equinox in 1994, I commented to a cousin that I felt like I'd over-slept and was sleep-walking through life—going through motions, but not feeling inspired. During and after my kundalini awakening, I sensed I was much more than just my physical body and egoic self—that I was also a higher level observer of many levels of self, capable of choosing how to emotionally and energetically respond to everything.

Experiencing the Mandela Effect can help us to gain a sense of how we exist in the form of consciousness, where we are essentially dreaming all the time. We can learn to awaken within the dream, like lucid dreamers, when we are consciously aware of witnessing Mandela Effects and finding meaning in signs, symbols, and synchronicity. We can see how our focus on how good things can get can bring healing to ourselves, to others, and to the Earth.

We can experience mental influences at a distance, and remarkable precognitive insights. We can benefit from mastering skills such as: conversing with the participatory cosmos, discovering who we are at ever higher levels of awareness, receiving intuitive guidance from future memories, playing with superpositions of states and dual memories to select favorite realities, choosing to stay entangled with those we most love, embodying favorite versions of ourselves, locking in preferred realities with focused intent, and aligning with universal energies to stay relaxed and energized.

REBIRTH & INITIATION

Emergence from the unknown—from the subconscious—can seem a lot like Jonah emerging from the whale, or like Luke Skywalker, Han Solo and Princess Leia emerging from the garbage compactor in *Star Wars*. The mythological significance of this motif is that of being reborn from the dynamic unknown depths of the subconscious. With the Mandela Effect, a sense of moving through the unknown can be quite profound and transformative.

Such transformation can have profound impact on our lives. Relationships may be lost or changed, and sometimes reformed after we've experienced the Mandela Effect. At this juncture, it's clear we can't return to living as we had before, and new relationships can form, with recognition of the significance of the Mandela Effect. Some relationships may come to a close, as interests change, and there suddenly seems to be a great deal less in common to talk about—as we now identify as Mandela Effect experiencers.

Thanks to support from association with mentors and experiencers, we can rise above stereotypical egoic identification. We can see ourselves as consciousness—as selves who are observers of thoughts, moods, feelings, and emotions. This idea is beautifully described in Michael Singer's book, *The Untethered Soul,* which helps people attain a wonderful state of awakening from within a dreamlike state. [9-3]

REVELATION & TRANSFORMATION

One revelatory key to inner transformation is to gain awareness that we are both mortal actors in a 3D material world, and simultaneously also powerful observers with freedom to choose how we respond to events. We see this theme play out in many books and movies, including Plato's parable of the cave. In Plato's parable, prisoners are chained and confined in a cave in such a way that they view shadows cast on the walls as being true reality. Only at such time as the prisoners are released from the cave is it possible for them to see that reality consists of much more than shadows on a cave wall. We see this same theme in the movie *The Matrix,* where Neo chooses to take the "red pill," and arrives at a higher level of conscious agency. In *The Never-Ending Story,* the main character, Bastien, becomes aware that he is both the reader and also the main character of the story, with everything depending on his choice to awaken and take action. At the point of knowing one is both actor and observer, awareness of true identity dawns, and the experience of self is forever transformed.

Some revelations and transformations involve unlearning what has been learned, and replacing old beliefs with new ones. The Mandela Effect qualifies as an Exceptional Human Experience (EHE), capable of initiating spiritual awakening. There is an advantage in focusing away from fears, as unique gifts, service and roles become more clear. Mandela Effects challenge us to let go of expectations that things will continue the way they seem to be, to ponder what is happening, and to ask why reality is shifting. We can then begin to see patterns in the shifts, and recognize that perhaps the answer to why reality shifts is simply, "Be Cause," and we are creators with quantum superpowers.

Revision of beliefs is key to the revelation and transformation part of the Hero's journey. It's essential for our growth and development to change how we live our lives to better align with our fuller sense of selves. We can see how much more energized, inspired, humble, and empowered we can be when aligned with the indomitable spirit of pure, unconditional divine love. As we welcome growth in our life, and choose to be our best possible self, we can experience being part of global evolutionary transformation. A large part of the reason for experiencing the Mandela Effect might just be to break free from hypnotic, prescribed personas living autopilot lives, and step into powerful, best possible versions of ourselves, as we remember who we truly are.

RETURN HOME MASTER OF MANY WORLDS

When we return home from our adventures with wisdom from our experiences, we can be respectfully received when we respect ourselves. By gently asking questions about what people remember and what matters to them, and by deeply listening to their experiences, it's possible to communicate more clearly and effectively.

It's possible to have inclusive, open-minded, and respectful conversations after experiencing the Mandela Effect. It can help to keep in mind that not everyone is willing or able to see the Mandela Effect, and not all experiences will match or be verifiable with others. Only some of those who are willing to see Mandela Effects will wonder how it could be possible that reality could change so radically, and many people will have little to no interest in the topic whatsoever.

When talking about the Mandela Effect and sharing experiences, it's good to keep in mind that we can disagree with "historical facts" and still be respectful of others. Rather than fixing our attention on trying to prove we are correct, it's helpful to be kind and compassionate, adopting a view of Copernicanism, where we might each be in our own separate Wigner Bubbles, discovering what we can share with others, rather than retreating into self-preoccupied solipsism.

Part of the Hero's return involves willingness to live in accordance with our highest values, while respecting and supporting the freedom and sovereign spirit of others who are doing the same. Together we can co-create a reality built on honesty, integrity, kindness, and love.

As we discover through the Mandela Effect that we are both Actors and Observers in our lives, we can bring this message to the world, by embodying higher-dimensional, more expansive versions of ourselves.

MANDELA EFFECT EXPERIENCERS CAN SAVE THE WORLD

If you've ever felt you're in the wrong world, consider the possibility that you might be here to help create a new one. Those of us who experience the Mandela Effect are gifted with experiencing a world being created and breathed into existence based on what we truly need, what we dearly love, and what we envision when asking questions such as, "How good can it get?"

It can feel disorienting to see that pretty much anything can change. Yet we can attain a newfound sense of faith in reality by acknowledging that when it comes to what matters most, we are just fine. We can also recognize numerous examples of miraculous positive changes occurring worldwide, that can provide further reassurance. Thanks to attitudes of openness and optimism, Mandela Effect experiencers can save the world, by acknowledging and activating innate, natural quantum superpowers.

AXIOGENESIS AND OPTIMALISM

The idea that we can trust that Nature selects what is best for all concerned, is the central thesis presented by philosopher Nicholas Rescher in his book, *Axiogenesis: An Essay in Metaphysical Optimalism*.[9-4] Addressing the subject of intelligent design from a logical, philosophical perspective, Rescher concludes that we live in an incredibly fine-tuned universe that is astonishingly well-suited for life. By focusing on Gottfried Wilhelm Leibniz's question, *"Why is there anything at all?"*, we can focus on obtaining an explanation for the emergence of a domain of existents from an existentially empty realm of a nonempty world. Such a quest for ultimate answers demands holism, and the basic tenet of Rescher's optimalism is that the best possible order exists, and the prevailing world order is the best that can be actualized, because that's for the best. Nature is the great optimizer, dealing in superlatives so that maximum effects are achieved with minimum energy.

We're seeing some remarkable miracles, such as with the healing of the hole in the ozone layer, the return of animals previously thought to be extinct, microbes that eat plastic, and improvements to our human physiology. Combined human positive focus of attention can help ensure that we see continuing evidence of miraculous shifts in reality.

ACTIVATE QUANTUM SUPERPOWERS

Physicist Erwin Schrödinger presented the idea that biological life is dependent upon quantum processes at a very fundamental level, suggesting that biological heredity is based upon the principle of "order from order," rather than "order from disorder" classical laws. [9-5] The deeper levels of meaning we can gain from the Mandela Effect have everything to do with remembering who we truly are. We are powerful, creative, imaginative beings, capable of viewing ourselves and our world from higher levels of observational awareness, and activating quantum superpowers.

One of the coolest things about entering the Quantum Age is that we can start utilizing quantum tools—and we don't have to wait for quantum computers, robots, or Artificial Intelligence! Thanks to Nature being fundamentally quantum, and our human capacity to be curious about quantum principles and ideas, we can begin enjoying quantum capabilities that each of us naturally has been given.

Quantum processes previously assumed to be relegated exclusively to the microscopic realm of quantum physics are starting to explain some of biology's biggest mysteries. And as one might expect in a young field such as quantum biology, evidence is still coming in, so we're seeing the first examples—such as photosynthesizing plants demonstrating quantum coherence and the ability to transport energy according to a quantum random walk—that can be presented with complete confidence at this time. Other biological examples, such as our sense of smell involving quantum teleportation, and the European robin's ability to navigate thanks to a quantum-entangled radical pair mechanism also appear to be fairly solid.

Some key concepts from quantum science dovetail nicely with natural gifts that can be explored when experiencing the Mandela Effect, as shown in Table 9-1. Here are ideas for how you can make practical use of these scientific principles—at play in quantum worlds of possibility!

Table 9-1:

Quantum Concepts & Natural Gifts

Quantum Concepts	Natural Gifts
Participatory Holographic Multiverse	Adopt beginner's mind; ask questions you need to live the answers to
Observer's Perspective effect	Get to know higher levels of self
Superposition of States	Play with future & multiple memories
Entanglement	Care for all you truly love
Quantum Jump	Show respect, courage, confidence
Quantum Zeno Effect (QZE)	Lock in desired realities by focusing attention on preferred outcomes
Elevate Energy States	Stay relaxed and energized
Quantum Eraser	Feel gratitude for desired pasts
Wigner Bubbles	Choose reality bubbles wisely
EPR Steering	Envision best for all concerned

Our natural gifts might seem simple, or unimportant, yet they are as essential as breathing. Masters of subtle Ki energies recognize that breath work is absolutely essential, and can be healing and life-saving.

NURTURE BEGINNER'S MIND IN OUR PARTICIPATORY, HOLOGRAPHIC MULTIVERSE

Some of the most astonishingly miraculous reality shifts can be experienced when we maintain a state of wonder and beginner's mind. There is delightful appeal in the innocence of young children, due to their purity of heart and lack of guile and egoic masks. Colors look brighter, worries and doubts vanish, and life shines with a brilliant quality of infinite possibility in the pure state of beginner's mind.

We can regain some of the magic and miracles possible in a natural state of innocence, or *shoshin*, as Zen Buddhists call beginner's mind. Modern day cognitive scientists find evidence of the value of embodying this beginner's state of mind with: openness, eagerness, lack of preconceptions, and genuine awe. Cognitive scientists inform us that we can increase our openness to new experiences. Openness is a "big five" fundamental personality trait, together with: extroversion, agreeableness, conscientiousness and neuroticism. People who are characterized as open-minded gain the benefit of making fortuitous connections, thanks to recognizing newly arriving possibilities in their peripheral awareness. By acknowledging new experiences, one can literally improve one's luck. Just as people experiencing Mandela Effects tend to notice them in their peripheral awareness, we can all make use of this tendency to witness what we'd most like to experience.

Beginner's luck comes from asking good questions, such as, *"What can I best do right now?"* while being open-minded and envisioning that of course it's possible to successfully accomplish whatever you set your heart and mind to do. Those who don't assume something is "impossible" are more likely to fortuitously live the answers to their favorite questions in our participatory holographic multiverse.

GET TO KNOW HIGHER LEVELS OF SELF
WITH OBSERVER'S PERSPECTIVE EFFECT

"Every man is a divinity in disguise,
a god playing the fool." —*Ralph Waldo Emerson*

When feeling aligned with higher self, we are inspired with wisdom, unconditional love, and compassion. There is a calm centeredness in this mindset, free from habitual behaviors or reflexive habits. Extraordinary synchronicities naturally unfold, and delightful personal Mandela Effects bring joyful blessing to everyday events.

Developing higher self awareness is a deeply fulfilling long-term activity. Practices such as regular meditation, yoga, qi gong, nature walks, and martial arts can be helpful in accessing higher levels of self. Some long-term changes we can enjoy when accessing our true eternal nature include: reduced anxiety, reduced depression, increased openness to life experiences, increased creativity, enhanced intuition, increased synchronicity, improved patience, improved fearlessness, and increased unconditional love for everyone.

It is possible to experience an observer's perspective effect in which our very choice of measurement—or the questions we ask silently or aloud—determines what we experience. This effect operates between us in all levels of our observational faculties, and with the participatory cosmos. How we choose what questions to ask and focus on can positively influence both our futures and our pasts.

PLAY WITH FUTURE AND MULTIPLE MEMORIES IN SUPERPOSITION OF STATES

Mandela Effect flip flops demonstrate the existence of parallel possible realities—similar to the quantum physics idea of superposition of states. One way to best enjoy superpositions of states is to be ready to recognize that experience when it arrives. We might, for example, notice Mandela Effect flip-flops between two or more different possibilities.

If you suspect you're experiencing a superposition of states, it's helpful to remain calm, and view non-preferred realities with relaxed detachment, as you might view food in a cafe's display case that you don't plan to order. By staying calm when I received multiple emails stating a package had been delivered to me that I'd not yet received, I ignored those, and focused on the lone email indicating the package was "out for delivery," which thankfully turned out to be the case.

I've experienced instantly gaining knowledge and skills I did not previously possess on some notable occasions. In the 1990s, I once was in the audience of a blues music performance, listening to musicians playing on a stage, when the drummer walked away, mid-song. I suddenly felt as if I was on the stage, playing those drums (an instrument I'd not played before), and I could feel that my hands and feet knew what to do, if I only had the nerve to jump up on stage and play. I reached a point where my discomfort of hearing no drums overrode my shyness and doubts—so I jumped up on stage, and had a surprisingly great time filling in on drums for the rest of that song, and a couple more. We can likely all access adjacent skills like this, without need for special technology, such as sometimes appears in films depicting navigation between multiple parallel universes, such as the movie, *Everything, Everywhere, All at Once*.

As you gain experience playing with superpositions of states, you can adopt a more joyful and grateful attitude—rather than feeling anxious or angry—while keeping desired outcomes in mind. If there is no specific desired outcome, you can enjoy witnessing adjacent possible realities, until the situation settles down, and one is selected. And of course you can optimize the entire process by asking with all your heart and soul, *"How good can it get?"*

CARE FOR ALL YOU TRULY LOVE
WITH QUANTUM ENTANGLEMENT

Like birds of a feather who flock together, we can benefit from feeling a sense of connection and emotional bonds with those we love. When we're entangled with people, animals, places, and events, we can feel a sense of being connected, regardless how far apart we may be in space and time. Whether your community consists of animals, plants, or people, you can enjoy opportunities every day to feel love, gratitude, and appreciation. I've enjoyed making some beautiful—and often surprising—connections with plants and animals when I was least expecting such experiences. Often this happens when I slow down and meditate outdoors, free from technological distractions.

ACT WITH SELF-RESPECT, COURAGE, & CONFIDENCE
TO MAKE QUANTUM JUMPS

Trust your memories, intuition, and feelings to guide you, and develop a sense of self-respect as you become the kind of person you most adore and admire. You are the hero in your life story, so be courageous! Our ability to make quantum jumps from one physical reality to another requires harmonious alignment between what we need, love, and are focused on (alignment within ourselves), together with a sufficient burst of energy to make a jump. The energy required is something like the rush of energy we might feel, for example, when making a jump for safety to save our lives.

LOCK IN DESIRED REALITIES
SUPPORTED BY QUANTUM ZENO EFFECT

"One cannot be prepared for something
while secretly believing it will not happen."
—*Nelson Mandela*

Once we see desired results along the lines of what we are most hoping for, it's helpful to make sure the wonderfully unfolding reality stays with us. And one of the best ways to harness the power of observation is to keep watching the proverbial quantum pot to lock in results you

223

prefer—and stop watching when less wonderful events are unfolding. Keeping a gratitude journal can be helpful for frequently observing how joyful and grateful you are to be attaining your most desired results. The key is to frequently feel grateful and blessed. It helps to also stop asking—even just in thoughts—any questions you'd rather not live the answers to, focusing as much as possible on how good it's been, how good it is, and how good it can possibly get.

STAY RELAXED & ENERGIZED
WITH ELEVATED ENERGY STATES

Just as you can discern the difference between a human being and a zombie that looks and behaves like a human, but has no human consciousness, you can tell when you feel like you're being fully yourself. Being yourself includes focusing on what matters to you, and caring about those you love. There's a sparkle in your eye and a spring in your step that is noticeably different from how you feel when going through habitual motions without feeling fully involved. And when you feel fully inspired, you are more likely to experience some of the most amazing Mandela Effects and quantum jumps. The experience of awe that we feel when encountering the unknown can bring us closer to staying in a state of reverence. Reverence and joy naturally arise when we come most fully into higher levels of consciousness. When feeling reverence and wonder, we can discover that every place is holy ground, and recognize inspiration everywhere. Children have freedom of personal expression, and are less concerned with whether their socks match or they say the right thing. Children are more likely to spontaneously do and say things that inspire them, while feeling relaxed and energized. Finding the essence of who you are can be found through play, meditation, and journaling.

FEEL GRATITUDE FOR DESIRED PASTS
WITH QUANTUM ERASER ASSISTANCE

The beauty of the quantum eraser is its ability to rewrite the past, through the power of selective observation. We can access the power of the quantum eraser by feeling gratitude for situations we don't yet have full details about, where we can participate in a kind of delayed choice. Such a delayed choice can operate in the sense that we are focusing our attention retroactively in a positive emotional way on a situation of as-yet-unknown outcome. Much like Gaelic speakers engaging in the imprecatory mode where we can feel hope—and gratitude—for past events, we can keep our minds and hearts open to positive developments in the future and the past.

CHOOSE REALITY BUBBLES WISELY
WITH COHERENCE IN PREFERRED WIGNER BUBBLES

The idea of "reality bubbles"or "Wigner bubbles" aligns with the view of Copernicanism, where individuals or groups consist of those who share encapsulated degrees of beliefs. There appears to exist a combination of free will and uniqueness of perspective, which taken together can serve to ensure that we can notice distinctly different realities than we've encountered before, as well as distinctly different realities than what have been experienced by others. Choosing wisely is all about making harmoniously aligned choices that satisfy all levels of self, from higher levels, to head, to heart, to gut.

ENVISION BEST FOR ALL CONCERNED
WITH EPR STEERING

Asking, *"How good can it get?!"* while focusing on how to utilize your unique gifts in service to the world at the highest level attainable can truly change the world in big, positive ways. It's possible to access the level of infinite, eternal consciousness where the most extraordinary reality shifts and quantum jumps occur—when feeling a genuine need to experience living in a world that is truly the best for all concerned.

Quantum EPR steering suggests that we can nonlocally affect or steer another system's states through local measurements, or observations. EPR steering can be helpful when observing news reports, so we keep in mind that even though facts appear to be one way, they can later be positively influenced to turn out completely differently. Nonlocal intercessory prayer is an example of how people work together to attain distant healing results for others. The net result can seem as if those being prayed for have jumped to better timelines. Some intention groups meet regularly in meditative prayer with focused attention, such as have been organized in Lynne McTaggart's Power of Eight groups.

TEST DRIVE
YOUR QUANTUM SUPERPOWER GIFTS

Now that you've explored how some of your natural gifts relate to various quantum processes, you can take your quantum superpowers out for a spin in something like a Cosmic back-to-the-future DeLorean!

In this quantum superpowered metaphor, you can steer where you are going thanks to your ability to stay focused on what you are visualizing, making sure that what you have in mind is also what is in your heart and gut. You can accelerate and move forward thanks to the emotional energies of love, gratitude, and reverence. Any time you'd like to lock in certain reality, you can hit the brakes with the Quantum Zeno Effect (QZE) by slowing down and focusing your attention. Whenever you wish to create at higher levels with more collaboration, connection and greater sense of spiritual expansion, you can shift into those higher gears with meditation, journaling from a higher level of self with *illeism*, and lucid dreaming.

Just like shamans described in John Perkins' book, *The World is as You Dream It*, those of us who witness Mandela Effects are capable of literally dreaming our world into existence. We can expand our High Sense Perception beyond the range of ordinary physical senses, harnessing intuitive insights and wisdom.

When we maintain inner harmony by being nonjudgmental with ourselves, and maintaining harmonious alignment between what we think, what we say, and what we do, we can experience the best results. Learning to employ positive self-talk is thus very important. Now is the time to speak with love to yourself, accepting whatever is happening as it unfolds, rather than feeling upset, anxious, or defeated.

We can learn to "cancel-cancel!" any destructively interfering thought-forms, regardless how much they insist that they belong. I feel blessed to have enjoyed some experiences playing with "quantum superpowers" on memorable occasions, such as when experiencing flip-flops. Just as my family helped our family dog switch out of possible realities where he developed cataracts, we can help each other switch out of almost any undesirable developing reality.

* * * * * * *

In the next chapter, *Humanity's Great Awakening*, we envision, imagine and explore the role of the Mandela Effect in humanity's Great Awakening.

~~~~~~~~~~~~~~~~~~~~~~~~~~~~~~~~~~~~~

EXERCISE:
ACCESS NATURAL GIFTS

What Quantum Superpowers would you most like to experience in your life? Imagine yourself engaging in daily life in this new state of mind, and how it might feel.

~~~~~~~~~~~~~~~~~~~~~~~~~~~~~~~~~~~~~

EXERCISE:
YOUR ACCIDENTAL HERO'S JOURNEY

Which part of the Mandela Effect Hero's Journey are you on at this time? How do you feel about your journey so far? What do you need?

~~~~~~~~~~~~~~~~~~~~~~~~~~~~~~~~~~~~~

EXERCISE:
APPRECIATE MENTORS

Which people provide you with a sense of peace, joy, calm, and inspiration? How can you best welcome and honor their presence in your life?

~~~~~~~~~~~~~~~~~~~~~~~~~~~~~~~~~~~~~

"The spirit won't go away.
It won't disappear or hide someplace.
It's everlasting."

—*Mabel McKay*

Chapter 10

HUMANITY'S GREAT AWAKENING

In 2019, the Scripps National Spelling Bee shocked everyone present when the announcer declared that the event had entered "uncharted territory." Eight finalists successfully endured 20 grueling rounds, at which point they were heralded as co-champions in the first eight-way tie in the spelling bee's history. [10-1] In this moment, the world caught a glimpse of win-win philosophy, with synergistic interactions ensuring benefits for all concerned.

Something extraordinary happens when people work together toward common goals, with win-win intentions. All transformative change may be rooted in this approach, with its power to transform even long-standing systems. Stephan A. Schwartz was astonished to discover how many successful social change movements in the USA began with, *"a few Quakers joining together in common intention."* Schwartz identified eight key success factors of the Quaker approach: shared intention, common goals, acceptance of possibly not seeing change in their lifetime, acceptance of lack of acknowledgment, fundamental equality, forswearing violence, making private selves consistent with public postures, and acting from integrity. [10-2] Taken together, these qualities eclipse egoic thinking, with a focus on living in accordance with harmonious cooperative intent.

Harmonious cooperation can work miracles when it honors deep wisdom. Tribal chairman, author, and former professor, Greg Sarris, shares Pomo basket weaver and medicine woman, Mabel McKay's inspirational life story and wise words regarding the everlasting nature of spirit in his book, *Weaving the Dream.* [10-3] Sarris has attained unprecedented co-management agreements between tribal and public lands, and coordinates scholarships and educational endowments, because, *"Our stories remind us to live responsibly on the Earth."* [10-4]

WISDOM EFFECT:
WE ARE THE ONES WE'VE BEEN WAITING FOR

The Mandela Effect (ME) can be viewed as a Wisdom Effect (WE) of greater shared awareness of interconnectedness in which humans become aware that we were created in the Creator's image to "be

229

cause" by responsibly acknowledging our divinity. *Homo sapiens* means "wise men," which can serve as a reminder to focus on what is best for our health and well-being, and the health and well-being of others. Cleverness can only take us so far—wisdom is far more valuable in terms of providing optimal guidance, especially in times of uncertainty and change. The Mandela Effect invites humanity to acknowledge our divinity, by rising above perceived human limitations, through ever-improving alignment with higher levels of conscious agency.

We can gain some clues and insights as to what this looks like from the Hopi Prophecy rock, a petroglyph rock carving situated near Oraibi, Arizona. The image on the Hopi Prophecy rock depicts how humans have undergone three previous ends of the world and are approaching a time of great transformation now. At times of great transformation, two possible ways of living are available. A materialist competitive win-lose path where peoples' heads are disconnected from their bodies (depicting lack of wisdom) will become bumpy and rough before it abruptly ends; while the win-win cooperative lifestyle of living in respectful harmony with love, strength, and balance will continue smoothly and steadily on. This pictograph also includes a line of connection between these two timelines, indicating that before the disconnected timeline gets too rough, people can change who they are, and become more loving, strong, balanced, and connected. Hopi advice for how to stay on track during end cycle times is simple: keep our hearts open to cooperating with others, and keep our heads open to humbly receiving intuitive and spiritual guidance. [10-5] [10-6]

TRANSCENDING EGO:
ALIGNMENT VERSUS CONTROL

Swedish physicist Carl Calleman's theory of waves of conscious creation provides us with insights into the way inspiration naturally flows through the cosmos. Calleman expected us to have received humanity's next great technology in October 2011—at a time when a tipping point of awareness of the Mandela Effect occurred. Calleman shows that the 9th wave introduces a self-transforming approach to life, allowing people opportunities to transcend egos, and identify as being Higher Self. Calleman states that ever since 2011,

> *"our will and how we create our lives is essentially a product of our unity with the Divine, and not of our ego."*

This time of Great Awakening invites us to let go of trying to control what's going on, and practice experiencing a relaxed state of flow.

Through observing thoughts and feelings, we can increase awareness of ourselves as being larger than any given role we play, or habitual behaviors. We can more gracefully live in harmonious balance without reacting to karmic stuff—and drama—such as obsessing that there are "bad guys" who did something terribly wrong. We can then focus on being aligned with what we most love and feel is life-affirming. We can more fully connect with sensing a hidden order in the cosmos, the Unity consciousness, the Absolute, the sense of ultimate quintessential goodness, and experience of a feeling of heaven on Earth. We can trust our intuition of subtle perceptions and feelings to guide us on this journey of experiencing love, kindness, wonder, and joy. We can bring healing and release angst associated with emotional and energetic shadow areas, ensuring we attract blessings, even in chaotic times.

Through meditation and prayer, we can connect with an internal sense of goodness and infinite, eternal love, and a state of inner purity. We can speak up about what matters most in ways that are respectfully cooperative. We can make sure that we are giving in highest service, and ask for what we need in return. We can be our own best friends, ensuring that our needs are a priority in the groups and communities we belong to. We can shepherd our internal resources and not over-extend our limits.

Learning to survive and thrive in community groups matters now, as the Mandela Effect encourages us to experience the answer to "How good can it get?" for all of us, rather than focusing on individual, limited self-interests. With acknowledgment that regardless of which of the various possible causes of the Mandela Effect are happening, ultimately consciousness—our consciousness—is involved, we can truly wish the best for all concerned, and find evidence that an optimal, collaborative, win-win future beckons to us now. This is the perfect time to collectively ask ourselves where we'd most like to go.

ENVISION INEVITABLE SUCCESS

In January 2012, Kerry Cassidy interviewed Bill Wood, aka B.V. Brockbrader, who said he'd worked with Project Looking Glass—a government program utilizing technology to view possible futures. Brockbrader stated:

> " … the project was shut down, because there was a problem when we approached 2012. I've heard it described a number of ways, but to my knowledge the problem is that the timelines converge on that point in time. And when you know enough about the Stargate projects and the Looking

Glass project to know how string theory works and how the possibility of possibilities works, and how making one choice over here doesn't necessarily mean that the other choice couldn't exist at the same time. But once you get your brain wrapped around this subject, you find out that at the end of 2012, an easy way to put it is that the choices that we make become less and less consequential to the future. And eventually we were pushed into this bottleneck of time, no matter which choice we make. And that's important to the people that had access to Looking Glass, because they would use Looking Glass knowing the choices that they would make in the future would pop up. The big mistake was coming up with the possibility of future. And when we started using a computer to say, well, if we make this choice, it's 79% possible that this scenario happens, and 23% possible that this scenario would happen. The understanding at the time was that was realistic. However, if you go down the road further, and free will continues to exercise itself on this game, that 79% possibility sometimes changes very, very fast. But if you look at the situation in a point of time, it seems very realistic that that's the greatest possibility. What happened was people—very smart people—began to figure out that something big was coming up. Something that made it so all the possibilities of all the future scenarios of any choice, any possibility, that was fed in and observed through Looking Glass inherently ended up in the same future, and no decision and no possibility changed past a certain point.... that's the big secret. It well coincides with December 21, 2012. All possible timelines lead to the same basic set of history in the future. What that is sends everybody that has all of the information that knows everything into a blind panic. The people that know everything about Looking Glass that have gotten all the reports and all the information—the elites of the world—probably figured out that that was the end of the game, when nothing could be manipulated beyond that point. .... I was called in and asked to solve this problem—this timeline contraction problem. And I eventually did my due diligence and did all the investigating, and basically only had one piece of information, and that was reinforcement; the computer is right. The timelines will contract down to some inevitable thing that you guys won't tell me about, so I can't help you. ... If I had to give it a name I would say it's the Awakening Process. It's an evolution of consciousness that cannot, will not, and no matter what decisions or possibilities are injected into the equation, eventually it all results down to us all learning the truth and becoming aware of this massive dam of lies that has been built, that keeps us from knowing massive volumes of information that we should otherwise possess." [10-7]

I independently heard from three people who emailed me after 2010, each one saying they used to work for the government in a time travel department, and they knew for sure that I didn't exist previously in their realities. They assured me, *"If you existed, I would have known."* I find

232

this fascinating for several reasons. First of all, apparently the government had a time travel department. Secondly, part of what they were employed to do was to be on the lookout for people such as myself writing about time shifts and reality shifts. And thirdly, they did not find me or my work during the time they were specifically tasked with researching and investigating these areas. In light of this information, it seems the time travel employees left those positions after 2010. This also happens to closely correspond with the time frame that Calleman anticipated happening in 2011.

WHERE IS THE MANDELA EFFECT TAKING US?

"We are entering a time where we are coming from what's in it for Me to what's in it for We."
—*Penny Kelly*

While we may not know for sure where humanity is headed as we travel between parallel possible worlds, this certainly seems to be a good time to ask ourselves where we'd most like to go. Mandela Effects indicate changes are taking place everywhere, including in: human anatomy, astronomy, biology, books, chemistry, geography, movies, products, and more. Some people notice personal changes in their bodies, with family members, and with neighbors, friends, co-workers and pets. People notice differences in what friends and family remember about past events, noting that some memories now seem impossible, yet the memories feel so authentic, detailed, and true. And Mandela Effect experiencers don't necessarily remember the same histories and events.

Asking why Mandela Effects occur can bring us to a place of more consciously playing with shifting realities, and quantum jumping from one reality to the next. We can become aware of the power our beliefs have over what physically happens in our lives. As I describe in my book, *Quantum Jumps,* we can see how the placebo effect has doubled in 30 years, and how people who have greater spiritual faith experience the most profound placebo benefits–feeling better when having taken some kind of "sham" treatment with no known medical efficacy. We are gaining insights into mind-matter interaction through the way our mind, body, and spirit work together. Results from embodied cognition research studies provide us with laboratory-tested ways to improve our levels of happiness, willpower, and intelligence by taking such simple actions as smiling, opening and closing our fists, and pointing to our head.

If there is no dividing line between the end of the classical physics realm and the beginning of the "quantum realm" as some scientists are recently contemplating, we begin to see what the Mandela Effect and reality shifts and quantum jumps are telling us, and where they may be taking us, too. [10-8] With no such dividing line, there may exist continuity from the realm of the very smallest subatomic particles and quantum wave functions to the realm of enormously huge star systems, galaxies, and constellations. Relativity may then apply both to spaceship travelers and quantum measurements and "quantum phenomena"—such as teleportation, tunneling, entanglement, and superposition of states—so that we can expect histories to change from time to time as consciousness makes various occasional leaps. Thanks to the quantum aspect, we can appreciate the profound power of observation, and the importance of aligned intentions, while operating from highest levels of conscious agency.

The Mandela Effect invites us to share what we have and what we know with one another, to courageously explore inner and outer worlds together. The Mandela Effect calls for us to overcome the divisiveness of viewing the world in terms of "Us" and "Them" with regard to religions, politics, ethnicity, gender, social status, or anything else—including levels and degrees of Mandela Effect experience, or who witnessed a given Mandela Effect "first." It is possible for each of us to adopt a graceful stance with regard to respecting everyone else's opinions and beliefs, without feeling "better than" or apart from others. The Mandela Effect shows us that facts can change, and people can disagree, yet still cooperate with and respect one another.

Some Mandela Effect-affected people, including YouTuber Evan Matraia (aka "Moneybags73"), have noticed that the letter "S" has been prominent in Mandela Effects of recent years, which seems to symbolize increasing awareness of how humanity is expanding and spiritually leveling up. Nora Yolles-Young spoke with me in a December 2019 interview about how she has noticed that people are becoming increasingly more aware of being psychic, and that as people can read one another's true intentions more readily, the kind of work she had done as a hypnotherapist is changing. It is easier to clear "shadow" issues of the subconscious, now that people are becoming much more open and transparent.

The Mandela Effect inspires us to envision how Cosmic Mind is constantly communicating with us, such that we can receive guidance, inspiration, and answers to questions. We can receive individualized, customized messages from such things as: radio shows, TV shows, license plates, our horoscopes, or the internet that can be spot-on and specifically tailored just for us at that exact place and time. This point of view of the universe conversing directly with us now has scientific merit, as expressed in the view of Copernicanism, by Eric Cavalcanti. This is *"the precise opposite of solipsism."* Copernicanism is *"the idea that no individual agent is at the center of the universe."* Cavalcanti also argues that taking this idea seriously requires a radical revision of the classical notion of "event" and by extension, "spacetime." [10-9]

SENSING NOWHEN AND THE ETERNAL NOW

"Eternity is not something that begins after you are dead.
It is going on all the time. We are in it now."
—*Charlotte Perkins Gilman*

The idea of accessing eternity can perhaps best be contemplated by imagining a point outside of time—an Archimedean "view from nowhen." We can access this in meditation by imagining slowing the flow of events down to a complete stop, and experiencing absolute stillness. The resulting sense of time can be felt as corresponding to spiritual, shamanic ideas of oneness in a present moment of infinite eternity. From such a vantage point, we might experience such things as: déjà vu, future memory, precognition, premonitions, intuitive hunches, synchronicity, and exceptional situational awareness. Déjà vu can gift us with a sense that we are experiencing events unfolding as we'd seen before. Future memories can be every bit as clear as memories of the past, such as the time I came across something I was about to purchase for someone while knowing "this shirt is his favorite," yet also that I was seeing it for the first time. Precognition, premonitions, and intuitive hunches can all feel like wonderful advice from our future selves. Synchronicity can feel like delightful orchestration of events that feel significantly related, yet that have no discernible causal connections. Exceptional situational awareness can provide us with a feeling of being aware of events unfolding from many perspectives, all at the same time.

While awareness of future choices and events may seem far-fetched, in the 1950s, Oxford philosopher Michael Dummett clarified the logical requirements by which reverse causality could transpire, such that future events could affect the past:

"... provided that the earlier event is detectable and that it can be detected without disturbing the circumstances under which later in time is claimed to cause the earlier event." [10-10]

In 1964, Yakir Aharonov worked with Peter Bergmann and Joel Lebowitz at Yeshiva University in New York to theorize that:

"The future can only affect the present if there is room to write its influence off as a mistake." [10-11]

One important thing to note here is that Nature loves plausible deniability, when future actions change past events! Another is that these theories that the future can affect the past are not mere thought experiments—they have been proven in a number of experimental studies.

Helmut Schmidt conducted successful retrocausality experiments in 1976, and Schmidt and Henry Stapp conducted experiments in 1993 in which experimental subjects successfully influenced previously recorded rates of radioactive decay. [10-12] Princeton Engineering Anomalies Research (PEAR) researchers Robert Jahn and Brenda Dunne discovered subjects influenced coin flips equally well either before or after the coin flip occurred. [10-13] In 2001, Leonard Leibovici published results from his double-blind study of more than 3,000 patients who had been prayed for some four to ten years later, who experienced considerably shorter hospital stays and shorter periods of fever. [10-14]

Awareness that we can employ causal agency both forward and backward in time can set us free from colonizing ourselves in linear time, as Glenn Parry points out in *Original Thinking*. We can reclaim an awareness of living in harmony with ourselves and one another, with confidence that we are not running out of time. One of my favorite works of fiction is *Momo*, by the author Michael Ende, who also wrote *The Neverending Story*. Momo is a little orphan girl living on the outskirts of town, who witnesses people enslaving themselves to time, and losing their quality of life—without realizing what is happening. Momo has perspective to see what is going on, because she lives in a world where people have time to talk to one another, listen to stories, and reflect. In Momo's world, people do meaningful work that they enjoy, that enriches their lives and community... until men in gray arrive. The men in gray promise people more time in the future if they save as much time as possible in the Time Bank by speeding up their work, cutting their social life, and destroying all joy, such that one observed:

"The odd thing was, no matter how much time he saved, he never had any to spare; in some mysterious way, it simply vanished. Imperceptibly at first, but then quite unmistakably, his days grew shorter and shorter." [10-15]

When we develop an intuitive appreciation that we are in the right place at the right time, and that we have everything we need, our relationship to time is positively transformed. We need not commodify time as a resource to save, bank, or store—and we can return to a more natural relationship with time, where an event and its time are one and the same. For example, to the Hopi, summer is the quality of being hot, since the time period is associated with how it is experienced. [10-16]

We can play a "no time" game, and let go of devices, watches, clocks, and any sense of having to be somewhere at some particular time. When living this way on a regular basis, we can experience increased synchronicity and meaningful coincidences free from habitual routines, and become more open to spontaneous invitations as they arise.

CONSCIOUS EVOLUTION WITH THE WISDOM EFFECT

*"Education is the most powerful weapon
which you can use to change the world."*
—*Nelson Mandela*

The Mandela Effect leads humanity out of assumptions of material realism, objectivity, and classical True-False logic into the realm of infinite possibilities. Thanks to the Mandela Effect, we see a dazzling array of possible futures, presents and pasts—with an invitation to seek meaning in the changes we observe. Many Mandela Effect experiencers witness extraordinary changes to reality in response to thoughts and feelings, which occur far beyond random chance. By viewing the Mandela Effect (ME) as a Wisdom Effect (WE), we can find evolutionary advantage in knowing that just because we *think* something is a certain way doesn't mean it will stay that way, or that others have experienced it that way.

Whereas there is an idea in manifestation with Law of Attraction that "thoughts become things," we might say with the Mandela Effect that we are the travelers to various realities and possible timelines via consciousness, which is our true identity. We tune into realities we select to experience, based on where we settle in terms of attentional, intentional, and imaginal "vibe." When we keep our vibes high, choosing gratitude, love, and reverence, we can travel to realities and events corresponding to our high vibe state of mind and being.

Acknowledging alternate histories, we see an explanation for why scientists are having a reproducability crisis, with difficulty replicating experimental results. We recognize a need to transform our legal systems as historians, psychologists, sociologists, anthropologists and biologists increasingly recognize alternate histories as a natural part of life. Medical professionals can officially acknowledge spontaneous remission as being a naturally occurring process, and doctors can encourage people to adopt "healthy beliefs" and states of mind facilitating quantum jumps for health and healing. Our views of unbiased observers and impartial judges will be forever transformed, as we appreciate how information can travel anywhere instantaneously, and how everyone and everything is interconnected.

EMBRACE OPTIMISM

Some Mandela Effects appear to represent a shift toward happiness, such as the gradually increasing smile of Leonardo da Vinci's *Mona Lisa*, and the appearance of the smiling crescent "Cheshire Moon" with a shape now fittingly positioned like a smile, rather than off to the side.

Perhaps the greatest gift of the Mandela Effect is how it can be experienced as a biofeedback mechanism for whatever vibratory frequency we choose. The Mandela Effect seems to be winking at us and saying, *"As you think, so you create!"* thereby providing us with opportunities to learn to stay in the vibes we most prefer. We can test this ourselves by living for a day in our natural mindset, whatever that might be—and then a few days later, seeing what happens when we raise our energetic and emotional mood.

How good can it get? We might be in heaven on Earth, and not fully realize it yet! Our new worldview can contain both visions of positive futures and positive pasts. It is possible that everyone is collaboratively co-creating thought-forms in such a way that invites order to arise out of chaos everywhere.

What we imagine and subsequently manifest is based on feelings of joyful anticipation and what we most love, rather than what we are most angry about, sad about, or afraid of. Rather than dwelling on conspiracy theories, we can adopt an attitude of envisioning *propiracy* theories, that are focused on such ideas as radical optimism. We can explore the concept of optimalism shared in Nicholas Rescher's *Axiogenesis*, and the optimistic science of philosopher Gottfried Wilhelm Leibniz. [10-17] As long as we keep asking questions such as, *"How good can it get?!"*, we are participating in creating a wonderful future.

238

LIVE MULTIDIMENSIONALLY

When viewing the cosmos through a nonlinear perspective, we can envision how we might all be time travelers. We can imagine connecting with possible future and past selves, gaining a higher-dimensional vantage point of knowing the fullness of who we are. The idea of moving from one level of dimensional consciousness to a higher level can be imagined when considering again, for example, the hypothetical residents of a two dimensional *Flatland*. [10-18] Such a 2D world would be completely flat, similar to a piece of paper, but without even any thickness of the page. Imaginary occupants of Flatland would only recognize and know shapes that fit on their two dimensional plane, such as: dots, lines, circles, and squares—but they would be unable to see what is hidden or concealed inside a circle or square. Only with evolved 3D vision could we have seemingly psychic superpowers to see inside a closed circle, by looking down from above to see what's drawn on a piece of paper, for example. And if we harness our 3D superpowers to push an object through Flatland, its residents would be amazed and mystified that objects could be so inexplicably transformed.

Enhanced levels of observational perspective grant us the benefit of higher dimensional thinking, as Leibniz originally suggested. Gottfried Wilhelm Leibniz, one of the inventors of Calculus, described consciousness in terms of levels of perceptions, writing:

> *"In order to be conscious of a 1st order perception, I must additionally possess a 2nd order perception of the 1st order perception."* [10-19]

Those of us experiencing the Mandela Effect might notice that we also appear to be experiencing a 3rd order perception, in which we remember things being different than "how they've always been." As Leibniz explained, we can increase our level of sentience through awareness of where and how we focus our attention. This awareness of the significance of the act of observation is also emphasized in the physics double slit experiment, where the very choice of observational device determines what is observed.

Feelings of lack and limitation can vanish when we connect with higher levels of spiritual consciousness, as Jacob Needleman pointed out. In a keynote speech for a Foundations of Mind conference, Needleman addressed the need to build a bridge between science of the brain and self-knowledge of the mind. He described beginning his investigation into consciousness as an atheist, and subsequently discovering powerful spiritual teachings. To materialist scientists demanding to know, "Where's the evidence?" Needleman responds,

> "There is nothing but evidence, but it is evident in the inner world, and requires a special attitude of humility, wish for truth, and a sense of the sacred."

Needleman continued that from this kind of inner empiricism, results will come! There are great benefits in recognizing that "*consciousness is a vast, vertical realm,*" so we may ask ourselves which level we need to be at as scientists in order to study consciousness. Feeling is an instrument of knowing, and

> "in moments of pure presence, we are who we wish most to be."

Needleman pointed out that "*We need a movement of the heart to study consciousness,*" adding, "*Only a virtuous human being can know reality.*" The words that resonated deep inside me with bell-like clarity weeks after the conference were Needleman's emphasis on the importance of:

> "... a special attitude of humility, wish for truth, and a sense of the sacred"

in order to be able to observe and comprehend the deepest truths about consciousness, that are otherwise kept concealed from human eyes and minds. These words inspire me with the work I do, assisting people to live authentic lives of inner harmony and balance. [10-20]

In any moments of doubt, we can find renewed inspiration in the words of Jesus Christ:

> "Very truly I tell you, whoever believes in me will do the works I have been doing, and they will do even greater things than these" (John 14:12) [10-21]

ALIGN WITH HIGHER CONSCIOUSNESS AND RISE ABOVE DRAMA AND CONTROL

I witness numerous positive personal Mandela Effects when I access higher levels of conscious awareness and maintain states of unconditional love, gratitude, and reverence. When I am most neutral and accepting of experiencing how good life can be, I witness miraculous moments. The key to success in alignment is to relax tendencies to exert excessive control, while welcoming new experiences that feel harmonious with the fullness of being. I feel humble making this connection consciously, with awareness of opening my heart to loving all of creation, and feeling great unconditional divine love in the stillness between thoughts.

By aligning with Divine Source and levels of higher self, we gain inspiration for becoming more open, flexible, observant, interactive, forgiving, in the flow, harmonious, kind, patient, honest, nurturing, and confident. Living in this state of alignment is quite different from egoic control, which is characterized by being closed, inflexible, unobservant, intolerant, impatient, imposing, and demanding.

By adopting a higher viewpoint above any drama, it's possible to remain in a state of free-flowing reverent optimism—with intuition from cosmic mind and a kind, compassionate heart. Extraordinary miracles are possible in this state of consciousness, with awareness that we share an entangled web of interconnected subjective realities.

The Karpman Drama Triangle depicts age-old roles of Victims, Saviors, and Aggressors, and patterns of behavior that are long familiar to us in our communities and families. Author Lynne Forrest describes this as victim consciousness, which she defines as *"the habit of thinking something outside of us is responsible for our happiness or unhappiness."* In actuality, the real key to thriving even amidst inevitable down times, is to realize that we are the ones who have the ultimate say in dwelling in victim consciousness. We can exit a state of victim consciousness by changing our beliefs. For example, if we feel that mistakes have been made, we can change this belief, and adopt new beliefs that there are no accidents, and there are no mistakes. We can thus move out of the victim-rescuer-persecutor mindset into observer-nurturer-asserter. It's important to show up with willingness to be open to new experiences, and willingness to learn from our emotions—including painful ones. Forrest shares ten universal laws of reality as we rise above the drama, including: (1) Cause and Effect, (2) Vibration, (3) Polarity, (4) Cycles, (5) Reflection, (6) Transmutation, (7) Verification, (8) Projection, (9) Intention, and (10) Alignment. [10-22]

A change in observer consciousness brings remarkable new opportunities, and beautiful unexpected gifts. At the exact moment I was writing the above paragraph that there are no accidents, and no mistakes, a friend emailed me a photograph of a fortune cookie fortune, *"Comfort zones are most often expanded through discomfort."* We can live in a state of graceful harmonious flow by changing our beliefs from fixating on what seems to be physically real and true, and recognizing that our conscious awareness extends far greater than that.

Awareness that we already exist in eternity—in the Archimedean "nowhen"—provides a deep foundation for thriving. When I interviewed Lynne Forrest for my *Living the Quantum Dream* podcast, she described how her mother taught her that:

241

What is eternal is true; all else is illusion.

By adopting the wisdom of this belief, we can recognize events in our lives as opportunities to develop our observer consciousness. We can thus gain a sense of the vastness of our true identity. [10-23]

CHOOSE THE REALITY WE NEED

Mandela Effects can be markers that other realities are nearby, so close at hand that sometimes we can witness them via flip-flops, where we see things that are one way and then another. By choosing our attitude, we can select the possible reality in which we reside. I've lived through this process, when my husband and daughter and I each in turn looked at our family dog's eyes clouding over, with each of us independently and together choosing that, *"sometimes it looks like our dog is getting cataracts, but then it clears up."* And it did. I've lived through this process when receiving four separate email notifications of a package delivery for an item I had yet to receive; three of which stated that the package had been delivered to my house, and one stating it was still on the way. Despite there being a three-to-one ratio of messages insisting I'd already received my non-existent package, I stayed calm and relaxed, with a strong sense of needing that package to still be on its way to me. And then I found out that it was. I went through this again at the 2019 Idaho conference, when the artwork that Kimberly-Lynn Hanson had arranged to ship from Canada to Sun Valley seemed to both be on its way, and not be on its way, and also already in the hotel! Kimberly received a call notifying her that the art had been delivered to the hotel the day before, according to their records—and there were two possible hotels. We agreed that we were choosing that the artwork was on its way, since we needed it to arrive. And it did!

REACH OUT AND CONNECT

The Mandela Effect invites us to step into our roles as cosmic service providers, having faith that through vast interconnections, we are all able to connect with and assist one another far more and far better than we may have ever imagined. Shane Robinson and Sherry Jagneaux point out that many logos and letters are extending and connecting now, seeming to suggest that as individuals we are reaching out and connecting more with others, on a much bigger scale than ever before. [10-24]

When we feel inspired to send a message to a friend, we are most likely making that connection precisely when our message can best be received. We can offer love, support, compassion and a sense of belonging that others can truly feel—sometimes even before the message is actually sent.

LOOK WITH SOFT EYES

I received an email reminder in September 2020 from author Carolyn North that there exists a more effective way to interact gently with the world that is, *"an old term used in animal tracking,"* of native American origin. "Soft eyes" employ a slightly unfocused gaze, ideal for taking in an entire scene all at once. "Hard eyes," in comparison, see nothing but a specific object of immediate interest, with too hard a gaze to receive all surrounding information. A soft eyes approach to life can develop the tremendous sensitivity of High Sense Perception skills. For example, a massage therapist was quite surprised one day to unexpectedly see the skeletal system of a client, *"... kind of like an X-ray. It was as if the bones suddenly lit up. I could see where the bones and muscles went together, and my hands just knew where to go."*

Looking with "soft eyes" can provide us additional information, in a form of visual deep listening, where we may not expect to see anything in particular, yet we can accept whatever we observe.

If reality is subjective rather than objective, then each of us has the ability to find, hold, and share possible realities with one another. If reality is subjective, then we are likely doing much more than we realize when we listen. We may literally be aligning and entangling with love—and with one another's unique worlds of possibilities.

LISTEN DEEPLY TO SENSE ADJACENT REALITIES

> *"Suppose we were able to share meanings freely without a compulsive urge to impose our view or conform to those of others and without distortion and self-deception. Would this not constitute a real revolution in culture?"*
> —David Bohm

One of the most transformational experiences in my life was attending SEED Graduate Institute's Language of Spirit Conferences with indigenous elders and scholars, scientists, and linguists. Many discussions involving nonlinear time took place at these Bohmian-inspired "Language of Spirit" dialogues and conferences, held in Albuquerque, New Mexico and led by Leroy Little Bear. Physicist

David Bohm attended the first Language of Spirit conference, where he shared his concept of dialogue as a way of communicating free from predefined purpose or agenda, that explores where a group of twenty or more people's conversation may go. The main idea is that when we stop defending our own ideas, we can more deeply listen to people saying things that we might otherwise block or feel insensitive to.

Language of Spirit dialogues and conferences created an opportunity for sharing ideas between indigenous scholars and elders, linguists and scientists from radically different walks of life. Leroy Little Bear selected which person would have the next opportunity to speak uninterrupted as long as that person felt inspired to continue. A special rapport developed in these dialogues, which became evident as various ideas wove their way in and out, creating a deeply transformative conversation.

These conferences mostly involved listening. A beautiful thing happened as the dialogue began with a sharing of individual views, and developed into heartfelt community. What amazed me most was noticing what happened inside me while listening to speakers who would often talk for long stretches of time. I noticed how my mind at first busied itself, sorting ideas and concepts into "facts" to be labeled, identified, and categorized. Even though Leroy Little Bear emphasized that the point of this dialogue is more about the *process* of dialogue than any particular statement, I initially caught my mind crafting a "perfect response." Within this struggle inside myself to simply listen, rather than prepare a reply, I eventually learned to relax, and be receptive to what was being said. My improvement in deep listening helped me see beyond limiting viewpoints. Best of all, this expanded state of mindfulness stayed with me long after the conference ended, inspiring me to more deeply connect with everyone in my life.

The practice of deep listening defies rational explanation, since dialogue participants experience something much more profound than simply a calming of "monkey mind." What happens is an experiential, whole-being awareness and embodiment of three shared great truths that American linguist Dan Moonhawk Alford summarized after the first gathering in 1992 in Kalamazoo, Michigan. Moonhawk noted three points of commonality were shared by quantum physicists and indigenous wisdom keepers:

(1) *Consciousness is primary and fundamental,*

(2) *Everything is vibration, and*

(3) *We are all interconnected.* [10-25]

Can you imagine how your life could improve if you could talk with yourself in the past? Can you imagine how your future self might be guiding you at this very moment? How would you feel if such communication through time was commonplace? These were some of the questions I shared in my presentation, *Listening to Future Selves, Reviewing Our Pasts,* at the August 2010 SEED Language of Spirit conference. One way of accessing all aspects of ourselves is by experiencing a timeless sense of Nowhen, in a transcendent meditative state of mind. This way of being in the world free from egoic control with its attendant fears, provides an optimal perspective from which we can experience an interconnected awareness of self and all that is.

Finding a state of deep listening and of deep conversation... without imposing old preexisting views or conforming to others' views can be deeply transformative. From such a place, it is possible to become aware of all aspects of oneself, including past, future, and parallel selves. It is also possible to move past a sense that it is always the other person who is wrong, prejudiced, or not listening.

BE IN SERVICE
WITH IDEAL PERSONALITY

We can feel most joyful and best aligned with spirit and consciousness when we are living lives of service. Blackfoot elder Leroy Little Bear contrasts aboriginal cultures with colonial cultures:

> *"Given the opportunity, Aboriginal cultures attempt to mould their members into ideal personalities. The ideal personality is one that shows strength both physically and spiritually. S/he is a person who is generous and shows kindness to all. S/he is a person who puts the group's needs ahead of the individual wants and desires. S/he is a person who, as a*

generalist, knows all the survival skills and has wisdom. S/he is a person steeped in spiritual and ritual knowledge. S/he is a person who, in view of all these expectations, goes about life and approaches 'all his/her relations' in a sea of friendship, easy-goingness, humour, and good feelings. S/he is a person who attempts to suppress inner feelings, anger, and disagreement with the group. S/he is a person who is expected to display bravery, hardiness, and strength against enemies and outsiders. S/he is a person who is adaptable and takes the world as it comes, without complaint."
[10-26]

The idea of embodying a unified sense of integrity within oneself and in cooperative relation with others provides a recognizable image of a natural leader who is valued and respected, and who operates at high levels of well-integrated and aligned conscious agency.

SHARE BRAINWAVES

We often assume each one of us exists as some kind of island, separate from others, with our own unique perspectives, biases, and motivations. Thoughtful awareness of our interactions with others can show us otherwise. The work of Walter J. Freeman shows how brains emit frequency vibrations, that taken together, are capable of sensing and setting the mood and feel of whatever groups we belong to. [10-27]

We can witness human participation with the hive mind of bees, such as shared by Dr. Larry Dossey, concerning the importance of "telling the bees." For hundreds of years, beekeepers have known the importance of telling the bees of any changes to the household through births, deaths, marriages and separations—or suffer the consequences. Bees who are not consciously communicated with are well known to abandon their hives, if, for example, a death in the family occurs of which they were not informed. Conversely, bees have been known to attend funeral services of people living on a property where their hive is located, almost as if they'd received an official invitation to pay their respects. [10-28]

Collective neuroscience is a relatively new and rapidly growing field of research. This means the experience of "being on the same wavelength" with another person is quite real, with researchers noting neurons in corresponding locations of different peoples' brains are firing at the same time in matching patterns. Such synchrony can occur between students with their teacher, and musicians and their audiences. Initial findings from this new field of study indicate that interbrain synchrony seems to facilitate the evolution of sociability, and appears to be highly beneficial. Relationship strength can be seen in stronger matches of brainwave patterns, such as is found between close friends, and outcomes of future interactions can be predicted by brainwave synchrony. Researchers are studying bat, mice, and human brainwave synchrony, and they have been surprised to find some of the highest levels of brainwave synchrony occurring between animals who were further apart in social status, and lower between animals who were closer in social rank. [10-29]

TOOLS IN THE WISDOM EFFECT TOOLKIT

The Mandela Effect is presenting us with natural gifts and corresponding quantum superpowers exactly when we need them. These gifts invite us to become adept at multidimensionally steering between possible realities in our lives. When we know that pretty much any problem can sometimes be miraculously solved, as if we arrived in a new timeline, we can better share these gifts with those we love.

One of the greatest gifts is that we may sometimes feel we have created key people in our lives. I had that distinct feeling when I found Dr. Yasunori Nomura, a theoretical physicist and prolific author of insightful research papers. Fellow IMEC board member, Shane Robinson, told me that he had seen Christopher Anatra arrive on the scene after having wished for a CEO to come forward and talk publicly about experiencing the Mandela Effect. It's a beautiful idea to consider that in a very real way, we are creating each other.

By trusting intuition, and taking a more active role in participating in imaginal realms, we can envision connecting with possible futures, with others whose presence we sense and feel, and with a sense of our best possible future selves reaching out to guide us.

LEARN WITH DREAMS – Dreams operating within the imaginal realm of possibility have access to all levels of conscious agency. Dreams can be remembered and reviewed upon waking, so we can learn from repeating patterns, symbolic content, and messages from the subconscious. You can tell yourself each time when going to sleep that you will remember your dreams, and keep a journal and pen nearby, so you can write your dreams down immediately after awakening. You can set an intention of waking up inside your dreams, with an affirmation mantra something along the lines of, *"Mind awake, body asleep,"* to experience a state of lucid awareness when sleeping. You can interpret your own dreams in a four-step process, to recognize messages and meaning unique to you.

(1)   Write a dream narrative sequence of events, including descriptions of place, time, environment, characters, and actions.

(2)   Rewrite the narrative by breaking it down into one individual meaningful component on each line.

(3)   Consider each individual key word, and free-associate what it means to you, writing down all the ideas that come to mind.

(4)   Read through all the words from step three, and rewrite the dream narrative in terms of the meanings that are revealed.

FIND MEANING – Journaling provides us with insights as to how we were feeling and what we were thinking about at the time when Mandela Effect changes occurred in our life. Journaling can also provide us with new ways to interpret and perceive past events. While we may not have direct command over events occurring in our lives, we can control how we interpret them. Writing about oneself in third person in the art of *illeism*, such as by writing your name in the journal entries, can provide increased awareness of patterns and meaning in your life events.

MAKE MEANINGFUL DECISIONS – We can exercise free will by activating our observer perspective consciousness in the present moment, with awareness that we are able to improvise and respond naturally, even outside past comfort zones of habitual manners of thought, speech, and behavior. When we make a choice to do something outside of our typical autopilot comfort zone, we gain the benefit of immediate alignment with preferred adjacent possible realities.

GET A SELF-IMAGE MAKEOVER – You can make a jump to your best possible life by envisioning how you would look, behave, think, feel, and speak—and then dress and groom yourself accordingly. You can start by literally changing your clothing, hairstyle, tone of voice, facial expressions, and mannerisms to better fit the person you intrinsically know your best possible self to be.

ASK GOOD QUESTIONS – When the questions you contemplate "on the inside" are aligned with the resonant fullness of your being, you will find answers to those questions. As physicist Henry Stapp explains, *"When you ask a question, Nature* *answers.*" Everything can speak to us in the ongoing conversation we have with the cosmos, and what is in our hearts and minds.

MEDITATION – This practice can help us clear away extraneous distractions of thoughts and feelings that are not truly part of this moment right now. You can try different types of meditation, and create your own by combining a favorite activity with a favorite focus of attention. Some different examples of this might be: singing while looking at flowers, or taking a walk while focusing on breathing deeply to your lower abdomen.

PRACTICE DEEP LISTENING – One of the best ways to be heard and seen by others is to listen. Deep listening is the best way to initiate true communication with everyone in your life, whether or not they recognize the Mandela Effect. Listening on a deep level is possible when refraining from interrupting those who are talking, and providing ample time and opportunity for people to fully express themselves without being interrupted or feeling defensive.

ENJOY SYNCHRONICITY – Noticing synchronicities of all sorts, including observing special numbers on digital clocks, license plates, and other places you might "randomly" happen to look can provide you with a marvelous sensation of living in constant two-way conversation with the cosmos. Pay attention to times when you are observing lots of synchronicities, and notice what sorts of things you are doing that might be influencing that.

MAKE PRE-EMPTIVE BLESSINGS – Know that you are a powerful creator who can bring balance, peace, love, and harmony to everything and everyone, including yourself in your future and past.

NURTURE SACRED SPACE – Just as you can help ensure your best health and vitality by tending to your personal energy field by: grounding, clearing energy cords, sealing your energy field, and conversing with angelic guides—so too can you clear and bless your home and workspace.

APPRECIATE REVHUMANISM – Recognize yourself and others as intrinsically divinely guided and connected to source. The Latin "rev" relates to ideas of reviving, regaining, recalling, and growing strong and young again. "Rev" also stands for reverence: a quality that inspires us to live up to our truly highest potential of seeing and respecting the inspiration, light, and consciousness that all of creation shares. Revhumanism invites humans to return to honoring the Earth as the source of life and sustenance. Revhumanism acknowledges that humanity is part of a living, growing cosmos, with everyone intrinsically having a transcendent divine nature. Revhumanism invites us to live true to our highest potential of embodying more wisdom than cleverness, more hope than cynicism, more humility than hubris, more empathy than apathy, and more reverence than insolence. Revhumanism represents a radical invitation for each of us to be the highest level embodiment of consciousness we wish to see in the world. When we see everyone and everything from the perspective of reverence, doors to adjacent possible realities open where there were no doors before. Revhumanism involves relating with reverence, humility, and empathy with others, with the Earth, and with the cosmos, inviting us to actuate high-level sovereign agency.

PRUNE TREE OF POSSIBLE YOUS – When choosing to block or prevent versions of you that you have absolutely no interest in experiencing, you can set intentions to self-limit, or "prune" all dead-end possible branches in alternate realities. You can envision a tree of possible selves going both forward and backward in time, and tend to it the way gardeners tend to *bonsai* plants, trimming possible timeline branches and roots.

STAY CLOSE TO GOD – Start and end each day with affirmational blessings to stay close to Divine Source / Creator / God, such as, *"I bless myself, my family, and my home,"* and *"How good can I get?"* and *"How God can I get?"*

THOUGHTS FOR CONSIDERATION

"Everything will be OK in the end;
if it's not okay, it's not the end."
—Fernando Sabino

The Mandela Effect is an Extraordinary Human Experience, providing humanity with deepening awareness of levels of conscious agency, subjective bubbles of reality, and clues to our extraordinary natural gifts and quantum superpowers. These gifts and superpowers are now arising from long dormancy to assist us in our most challenging and uncertain times. The Mandela Effect provides humanity with a new kind of collective consciousness technology—exactly when we need it most—to help us thrive in the next stage of human awakening and personal and collective evolution.

Mandela Effect experiencers are adventurers on an often accidental Hero's Journey. If things ever seem bad, it's helpful to remind ourselves that our story is not yet over, and we've just encountered an "interesting" part of the tale. We can ask to see the gifts in every situation, and sometimes receive surprising answers illuminating silver linings, pearls, and gems everywhere.

As we become conscious agents operating at higher levels of intentional and attentional awareness, we can learn to cooperate harmoniously and create wonderful spontaneous collective reality shifts. We'll do well to keep asking, with a burning, heartfelt sense of truly needing to know: *"How good can it get?!"* for ourselves and all our relations.

The invitation we are receiving from consciousness and the cosmos at this time of great awakening is that causality is not unidirectional in time and space; it is based on relationships. Consciousness is the fundamental foundation for all that exists, and we can harmoniously co-exist with respect, honesty, integrity, and love.

* * * * * * *

The next chapter presents, *Mandela Effect Questions and Answers*, covering a broad spectrum of ideas related to this topic and the community.

~~~~~~

EXERCISE:
PRACTICE SOFT EYES AND DEEP LISTENING

Next time you spend time with a friend, colleague, or family member, practice the expansive intuitive art of looking with "soft eyes" and listening deeply, without attempting to prepare your next words or response. Take in the holistic fullness of sharing space and time with them. Afterward, write down aspects of this time together where you observed something special that you most likely wouldn't have noticed with "hard eyes" or shallow listening.

~~~~~~

EXERCISE:
PLAY A NO-TIME GAME

Let go of devices, watches, clocks, and any sense of prescheduled activities. Find your own natural rhythm of eating, sleeping, working, and playing—and stay open to spontaneous invitations as they arrive.

~~~~~~

EXERCISE:
PRUNE TREE OF POSSIBLE YOUS

You can move toward becoming your best possible future self, or ideal personality, thanks to viewing and pruning possible past and possible future versions of you. You can imagine pruning out those deselected possible timelines, like trimming roots and branches on a *bonsai* tree. You can start by identifying possible versions of you that you'd prefer don't cause hardships for others, selecting whatever specific qualities are attributes you'd like to select and deselect in yourself.

~~~~~~

~~~~~~

EXERCISE:
EMPOWERED WITH WHAT YOU CAN DO

Make a list of what you can realistically accomplish today, making sure to include what you know is best for you to tend to. Release all the things you can't do today as being beyond your personal control for now, and have faith that Creator/Source/the Universe/God has got this. Take deep breaths, and relax, knowing you're always doing great when you've done your best. Ask "How good can today get?" with all your heart, and see how well today can go.

~~~~~~

"The important thing is not to stop questioning.
Curiosity has its own reason for existing.
One cannot help but be in awe when he contemplates
the mysteries of eternity, of life,
of the marvelous structure of reality.
It is enough if one tries merely to comprehend
a little of this mystery every day."

—*Albert Einstein*

Chapter 11

MANDELA EFFECT Q&A

Q: Is everyone in the entire world affected by Mandela Effect and don't know it, or is there just a special group of people who are affected and able to see this?

A: This question was addressed to the panel of speakers at the 2019 West Coast Mandela Effect conference, and all five speakers—Chris Anatra, Kimberly-Lynn Hanson, Jerry Hicks, Cynthia Sue Larson, and Shane Robinson—agreed we are all affected. This appears to be the way Nature works, and the field of Quantum Biology is beginning to show us that our very existence depends on quantum phenomena occurring on the macroscopic scale—which we also recognize as the Mandela Effect, reality shifts, or quantum jumps. [11-1]

_____

Q: Why do some people not seem to notice Mandela Effects?

A: Some people are ready to see the Mandela Effect, and others are not. Mandela Effect changes occur in the periphery of our awareness —rather than at the focal point of our attention—and they can seem relatively inconsequential, and not directly impacting our daily lives. People are more likely to witness physiological changes if they are not medical experts, dealing with the human body constantly. Doctors are more likely to believe "the heart has always been in the center of the chest," and "the kidneys have always been where they are."

_____

Q: How widespread is the Mandela Effect? Is it happening outside of the USA?

A: Reality shifts have been reported all around the world since the 1990s, when such reports were first collected. Documented reports from Denmark and the United Kingdom show the media has covered news reports of mass "mis-remembering" in confusion regarding what happened in a popular Danish TV series, or what happened to a prominently featured dinosaur at the natural history section of the Bolton Museum in England. People witnessed buildings and roads

255

transform, appear, and disappear around the world. IMEC discussed Mandela Effects in Brazil in a special episode. [11-2]

Q: Why do most Mandela Effect changes seem small and relatively insignificant, in the scheme of things?

A: If the underlying cause of the Mandela Effect is consciousness, and some higher levels of consciousness are inviting us to open our minds beyond the limitations of linear, two-dimensional concepts of time, then we can imagine that such an awakening would be gentle.

Q: Why do we notice some Mandela Effects flip-flop, or change again—such as eye color of the actors on Gilligan's Island?

A: Quantum Businessman Chris Anatra created YouTube videos asking why the eye colors of *Gilligan's Island* characters would flip-flop—and change back to something else, after having turned green. Sometimes when we observe Mandela Effect flip-flops, we witness dynamics between the way the Quantum Zeno Effect (QZE) can effectively "lock in" some changes, while quantum entanglement and coherence can help explain the tug-of-war that ensues when groups with different intentions are at odds, with destructive interference.

Q: Is everything already here and created, and we shift into it?

A: We can see from various interpretations of quantum physics that all possible realities can be viewed as already existing in original thinking, or the Dream Time. This corresponds beautifully with the ideas of indigenous cultures who appreciate that rational thought is a subset of a greater awareness, and benefits of restored balance are possible when we allow for larger awareness to inspire us. [11-3]

Q: Why are we having vehicle issues? I am having one vehicle problem after another.

A: We can view any kind of recurring "problem" as a dream message, which we can read like a waking dream. For problems involving recurring issues with transportation, we can investigate what the vehicles represent, and glean insights regarding how we are choosing to move through our life. Our forms of transportation can represent our ability to feel we can accomplish our goals with some degree of achieving what we set out to accomplish. Real life experiences can be interpreted with four step dream analysis, to find clues and meaning.

Q: What advice do you have for people who want confirmation that this is a real phenomenon?

A: Those seeking evidence of Mandela Effects, also known as reality shifts, can find reassurance in others who remember similar alternate histories—indicating that we are most likely living in not one, but rather many, worlds. In addition to finding support from others with similar memories, we can also find further supporting evidence that reality shifts are real in flip-flopping Mandela Effects. Flip-flopping reality shifts provide us with the opportunity to see something change back and forth from one state to another, each time seeming as if it's always been that way. Noticing such shifts can be especially helpful for providing someone who is unsure whether their mind is playing tricks on them, to know that a Mandela Effect has actually occurred.

Q: How do you think differently now that you know facts can—and often do—change?

A: When I began encountering undeniably large numbers of reality shifts in 1994, I wanted to ignore this phenomenon. I hoped I was imagining things, and that I had **not** just witnessed a woman vanish in front of me, or a coat change from one kind of fabric to another, or a massive concrete sundial sculpture appear where it had not been before. It seems to be human nature to first attempt to fit our experiences into established belief structures. I found it was easier to discuss this subject with young children, who are generally more open to directly experiencing the world as it is, without attempting to map some theory or belief structure on top of their observations. Once I gained a sense of familiarity with reality shifts, I felt more at ease when I noticed something change. I felt thrilled and awe-struck, like when I face any natural wonder of the world (such as an active volcano, a beautiful sunset, or a mother bird feeding her chicks in the nest). I love the sense of awe that reality shifts bring me, and I feel much more alive and inspired when I see reality shifting around me. I have a sense of hope—that no matter how things may seem, they can always change for the better. I've become much more of an optimist since I came to know that facts can—and often do—change.

Q: How can I feel less uncomfortable with the idea that things aren't as real as I thought they were?

A: You can adapt to moving through life with this new viewpoint, with appreciation that this accompanies the evolution of human consciousness. It might help to remind yourself that the kinds of changes we typically experience are mostly rather small and benign, so you can learn to relax and eventually get into a state of excitement and joyful anticipation about how good life can get.

––––––––––

Q: Might it be possible that the Mandela Effect could be explained by a combination of accepting the possibility of the reality of time travel and that we exist in a simulation?

A: These two theories are incredibly popular, and many people don't realize that the active components of each of these theories are part of existing constructs. Namely, quantum physics already shows us that future observations influence past events, and we already know from perception experts such as Dr. Donald Hoffman that our physical senses never show us the truth of what is "out there," but rather provide a simulacrum, or simulated semblance. So a dreamlike simulated reality involving time travel can be recognized as part of the natural way that levels of consciousness participate in mind-matter interaction.

––––––––––

Q: Why do some people experience more reality shifts than others? Are they more enlightened, or have a higher energy vibration that somehow affects multidimensional spacetime, or somehow affects those people on a quantum level? How can I improve my chances of more actively experiencing reality shifts?

A: There likely is more than one reason why some people notice and observe more reality shifts than others. There are some characteristics correlated with reality shift and Mandela Effect experiencers. Perhaps most notably, people who experience reality shifts consider themselves to be natural empaths, or highly empathic. On the Myers-Briggs personality test, experiencers are often Intuitive Feelers. One's ability to change perspective quickly may play a pivotal role in one's ability to experience reality shifts, and this could explain why empaths and Intuitive Feelers observe reality shifts more often than others. It is possible to enhance one's empathy, such as by reading works of fiction where you glean insights into how another thinks, or by conversing with people who seem different than you, and do your best to deeply listen and see the world from their point of view. Energizing yourself can help, and you can boost your internal energy with breathwork, Qi Gong, or yoga. You can also commune with nature, and see the world through the eyes of a squirrel, a bird, a cloud, or a tree.

Q: The idea of the Mandela Effect is interesting, yet also a bit troubling. I don't believe in solipsism, but when reality shifts involve other people, it is a bit difficult to comprehend. Are these other people just actors in our personal life drama?

A: I prefer the idea of Copernicanism to solipsism; Copernicanism is a term coined by physicist Eric Cavalcanti, to describe personalist views of quantum states. Through such a concept, combined with Cavalcanti's concept of *Wigner Bubbles*, we find ourselves in a cosmos where each conscious agent's choice of observational perspective grants them a uniquely personal set of facts and events—some of which may be shared with others who are in the same or similar reality bubbles. While some other people might sometimes seem like non-player characters at times, you can listen to people and experience the world a bit from their point of view.

Q: I experienced a number of déjà vu, starting as a child through my 30s, and a series of Mandela Effects in the 1980s. Over the past several years I am not aware of these experiences occurring. It could be I am not paying attention. However, many other subtle and not-so-subtle synchronicity are being experienced. Does this have any significance that I am not wise enough to understand? Could previous experiences have been something we all need to go through to learn from to move to another level?

A: I've heard similar things from people over the years—that while people had at some point experienced déjà vu (sometimes quite extensively), it happened earlier in their lives, and was subsequently less often experienced. Déjà vu is a remarkably common experience that an estimated 70% of the population experiences—yet for some reason, people most often notice it between 15 and 25 years of age. I've experienced déjà vu in my life into my 30s, while I was also noticing reality shifts and personal Mandela Effects. When I've had déjà vu, I felt that what I remembered having happened was something I've dreamed about in the past. When I've had the kind of Mandela Effect in which a sequence of events exactly repeats, something like a time loop, or what happened in the movie *The Matrix* that Neo observed—

that is quite different from déjà vu. These time loops are not simply a feeling that I've seen something happen before as if in a dream, but rather a genuine knowing that I have personally witnessed these exact events happening in nearly identical fashion. It's possible that we can first encounter the more typical déjà vu experiences, and then once we're aware of the Mandela Effect, we can sometimes witness time loops where we know for sure we've seen these events happen before. Some Mandela Effect-affected individuals are noticing, for example, that a particular YouTube video is released supposedly for the first time, yet many subscribers remember that this video came out earlier. With regard to your still experiencing synchronicity, I consider synchronicity to be very much a part of reality shifts and personal Mandela Effects, as they are evidence of extraordinary alignment between what you are mentally focusing on and what is happening "out there" in the physical world. This kind of mind-matter interaction is a hallmark of all reality shifts, Mandela Effects, and quantum jumps, and it's clearly evident in synchronicity, too. As far as "levels" of reality shifts and Mandela Effects go, it seems our beliefs and assumptions play a large part in which kinds of reality shifts we most likely experience. In cultures where, for example, there are words describing that thing that happens when people are heard before they actually arrive, such things are observed happening more often. Now that humanity is gaining a number of words to describe reality shifts, quantum jumps and the Mandela Effect, we will more likely experience more of this phenomena. As we open our minds to what may be possible, we can benefit more from things such as the placebo effect, and constructive shifts in ecology, world peace, cleaner environment, return of previously extinct species, improvements to human physiology, and much more.

Q: How much of our experience is shared? Do our realities split, based on who observes it? If you observe a quark in one position and I observe it in another, certainly we're both correct, but we can't live on the same plane of reality at the same time, with different observations. Does that mean that the only person that actually exists in my reality is me? I'm only capable of observing one instance of me at a time, so then who plays me in their realities? Do I even exist in their realities?

A: One thing we're witnessing now is that not everyone remembers the same facts the same way; this phenomenon is known as the Mandela Effect, and growing numbers of people are noticing that they do not actually remember everything exactly the same way as others. Some changes are being observed in historical facts, such as military events, geographic positions of land masses, and human anatomical physiology. The types of changes being witnessed suggest that even for those who are remembering facts differently, what they remember differently does not always match 100% with others. So we quite literally are witnessing some behind-the-scenes macroscopic quantum phenomena that likely provides us with our fundamental physical reality. In order to understand how your sense of reality interacts with others, the first step is to come into greater awareness with who "you" truly are. We have been taught in modern educational systems that each of us is our body, such that we seek material descriptions and definitions of our mind in the neural networks in our brains, for example. Even for those who seek to determine more of a sense of mental field, we are not absolutely clear on where one of us truly extends, and how and where our brainwaves and energies interact with others. Again, we glean insights and clues from observing the Mandela Effect in this regard. For those who are said to "take the download" or undergo "the Mr. Smith effect," it's fascinating to see someone start to describe the way they remember something, such as the physical location of their heart or their kidneys in their body, and then sometimes when they find out what they remember is not how these things "have always been," you can literally see that person suddenly remember the way those organs have now "always been." This fascinating process provides us with insights into how collective consciousness operates, such that we can witness the moment when someone's consciousness realizes it's like a lone bird that's strayed from it's flock, and how instantaneously it can move to rejoin that Wigner Bubble group of coherent consciousness.

Q: If you "leave" one universe and jump to an alternative one, is there a YOU still in the one you left? And if there is a YOU in the one you're jumping to, where does THAT you go to?

A: The biggest assumption and greatest opportunity in all of science has to do with assuming we know who "I" am, and we assume we know who "you" are. Thanks to the bias of the past four centuries of materialist, reductionist science, most of us tend to assume that our identity must be based in our physical form. Our modern governments thus fund extravagant, expensive neuroscience "brain mapping" projects consuming millions of dollars and euros, that do not deliver expected results. Our true identity is not some measurable thing that

261

can be seen in neural or brain maps—it resides in something we might never measure with physical devices inside our physical bodies. Our true identity is best considered as being composed of pure consciousness. This identity as consciousness is thus the "me" or "you" who is making any given quantum jump. In the "form" of consciousness, we jump from possible physical reality to possible physical reality. This appears to be the way of Nature, as we see in the quantum random walks selected with great efficiency in photosynthesizing plants. We also select optimal routes and paths that we, as consciousness, choose to take, much the way we might choose our path through a maze, where the path we select determines the life events and physical realities that we experience. We bring physical realities "alive" by gracing any physical reality with our conscious, energetic presence.

Q: I am finding family is bringing up things we have supposedly done that I have never done. It is trippy.

A: Wow, and yes, it is trippy when that happens! I suspect more people would realize they are affected by the Mandela Effect if they just took some time to share and compare old family stories.

Q: When we think of the discovery that plants perform photosynthesis the quantum way, is the plant the one that "explores" all the paths and then chooses the optimal one?

A: This matter of "conscious agent" is essential to many quantum interpretations, yet there is not yet total agreement on who conscious agents might be. My own personal view of this is that each of us is operating on more than just one level of conscious agency. We have levels of self corresponding with levels of perception, perceptual awareness, and awareness of those awarenesses. Each higher order level of conscious self can activate a higher state of consciousness.

Q: I have a question about the Quantum Zeno effect, about how to put a brake to a shift, to lock a quantum system, in order to maintain a desired result. How can we continuously or repetitively "measure or observe" a shift while going on with our lives?

A: Yes, it's a balancing act to continuously or repetitively lock a desired reality system in place, while still going about daily business. One simple way to continue observing successful reality is to practice gratitude journaling on a regular, daily basis. By genuinely feeling grateful for all the aspects of this reality you most enjoy, you can help to ensure that they stay actively present in your life. While we know how to observe and pay attention, there are ways to improve and expand our abilities. Various types of meditation provide tools by which we can ignore distractions from our internal thoughts and feelings, and also external events. The quantum zeno effect (QZE) provides an observer with the ability to continually observe a given state such that it can be "locked in" to a state of stasis, making sure the proverbial "watched pot" never boils. We can develop awareness of how we are asking yes or no questions, as proposed by physicist John Archibald Wheeler. Wheeler coined the term, black hole, and influenced many of the world's top physicists, including Albert Einstein. Wheeler proposed that we live in a participatory cosmos, where the way we make observations influences the reality we subsequently observe. Wheeler described how physical reality "it" comes from informational "bit"—physical reality arises from information—as our questions create the world. We see some evidence of this in experiments demonstrating how two observers at the same place and time can witness completely different subjective realities. Physicist Henry Stapp also adopts this way of viewing the importance of free will and conscious choice in how we observe and what we subsequently perceive as being pivotal for our experience of reality, and has written books and articles about this. Because our questions can so thoroughly influence reality, we can learn to individually utilize this influence by more carefully tending to how and what we pay attention to.

*"If you haven't found something strange
during the day,
it hasn't been much of a day."*

—John Archibald Wheeler

REFERENCES

CHAPTER ONE:
INTRODUCTION TO THE MANDELA EFFECT

[1-1] Bell, Art. *Coast-to-Coast AM*. April 16, 2001.

[1-2] Dickey, Jack. "Insomniac Radio King Art Bell Reclaims His Crown." *Time*. September 23, 2013.

[1-3] Bell, Art. "Cynthia Larson – Reality Shifting." *Midnight in the Desert*. October 15, 2015. https://youtu.be/gGXGbHjBL1E

[1-4] Broome, Fiona. "Mandela Effect Turning Point? Back to Science and FUN speculation!" Fiona Broome blog. July 24, 2022. https://fionabroome.com/mandela-effect-turning-point/

[1-5] Colts, Eileen. "IMEC 2020 Presentation." *International Mandela Effect Conference*. 22 Jun 2020. https://youtu.be/yA8dN_RTQbM

[1-6] Colts, Eileen; Pinto, PauloM.; Robinson, Shane C.; and Vannessa VA. *Mandela Effect: Friend or Foe?* 11:11 Publishing House. Estero, Florida, USA. 2019. Page 17.

[1-7] "The 21st Century News." *Head of the Class*. Season 2, Episode 21. 4 May 1988.

[1-8] Atwater, P. M. H. *Future memory*. Hampton Roads Publishing, 2013.

[1-9] Larson, Cynthia Sue. *Reality Shifts: When Consciousness Changes the Physical World*. RealityShifters, 2012.

[1-10] Larson, Cynthia Sue. "Your Stories." RealityShifters. http://www.realityshifters.com/pages/yourstories.html

[1-11] LaFlamme, Mark. Street Talk: The Mandela Effect is freaking me out. Sun Journal. 28 Jun 2016. https://www.sunjournal.com/2016/06/28/street-talk-mandela-effect-freaking/

[1-12] Puschmann, Karl. Berenstein or Berenstain? The riddle making book lovers mad. New Zealand Herald. 21 Aug 2015. https://www.nzherald.co.nz/entertainment/news/article.cfm?c_id=1501119&objectid=11500292

[1-13] Siebert, Tom. Technology and memory down the rabbit hole. San Diego City Beat. San Diego, California, USA. 25 Jul 2016. http://sdcitybeat.com/culture/all-things-tech/technology-memory-rabbit-hole/

[1-14] Anatra, Christopher. Quantum Physics, the Mandela Effect and perceived changes to your NECS entrée data. 25 Jun 2019. https://youtu.be/XUPxDLMCUKM

[1-15] Harris, Matthew. A Discussion with CEO Chris Anatra: What the Mandela Effect has meant for his business. Geek Insider. 10 Jul 2019. https://www.geekinsider.com/a-discussion-with-ceo-chris-anatra-what-the-mandela-effect-has-meant-for-his-business/

[1-16] Chalmers, D.J. *The Conscious Mind: In Search of a Fundamental Theory.* Oxford: Oxford University Press, 1996.

[1-17] Frank, A., Gleiser M., and Thompson E. (2019) The Blind Spot. Aeon Magazine.

[1-18] Goswami, Amit. *The self-aware universe: How consciousness creates the material world.* Penguin, 1995.

[1-19] Emerging Technology from the arXiv, MIT Technology Review. 12 Mar 2019. https://www.technologyreview.com/s/613092/a-quantum-experiment-suggests-theres-no-such-thing-as-objective-reality/

[1-20] YouGov Survey: Mandela Effect. Interviewing Dates Aug 23-26, 2022.

[1-21] Larson, Cynthia. How Do You Shift Reality? The First Ever Reality Shifting Survey Results Are In! *RealityShifters.* April 2000. http://realityshifters.com/pages/research/apr2000.html

[1-22] Larson, Cynthia Sue. *Quantum Jumps: An Extraordinary Science of Happiness and Prosperity.* RealityShifters, 2013.

[1-23] Tweed, Carter. Mandela Effect Analytics. 21 Nov 2019. https://www.alternatememories.com/analytics

[1-24] Larson, Cynthia. *RealityShifters Monthly Ezine.* October 1999 – 2020. http://www.realityshifters.com/pages/news.html

[1-25] Stine. Remembering Danish "Matador" TV Show Scene that Doesn't Exist. RealityShifters: Keep a Tidy Mind. Issue 227. August 2018. http://www.realityshifters.com/pages/archives/aug18.html

[1-26] Matador TV series. 1978-1982. https://www.imdb.com/title/tt0077051/

[1-27] The Mandela Effect in Brazil. International Mandela Effect Conference (IMEC) Open Tables. Season 1, Episode 4. 26 May 2021. https://www.youtube.com/live/lCzm5psm3Yo

CHAPTER TWO:
MANDELA EFFECT SOCIETY

[2-1] Haidt, Jonathan. *The righteous mind: Why good people are divided by politics and religion.*

[2-2] French, Aaron. "The Mandela Effect and New Memory." *Correspondences* 6, no. 2 (2019).

[2-3] Tart, Charles. "The archives of scientists' transcendent experiences (TASTE)." *Journal of Near-death Studies* 19, no. 2 (2000): 132-134.

[2-4] Tart, Charles T. The Archives of Scientists' Transcendent Experiences (TASTE) http://www.issc-taste.org/main/introduction.shtml

[2-5] Larson, Cynthia. *RealityShifters Monthly Ezine.* October 1999 – 2020. http://www.realityshifters.com/pages/news.html

[2-6] Cicarino, Carla, Christopher Anatra, Jerry Hicks, and Cynthia Sue Larson. The Mandela Effect in Brazil. International Mandela Effect Conference (IMEC) Open Tables. Season 1, Episode 4. 26 May 2021. https://www.youtube.com/live/lCzm5psm3Yo

[2-7] 101 SRPHD Truth Exposed Channel. Mandela Effect: Carl Sagan Remembers the Sagittarius Arm/Carina Cygnus Spiral Arm. 1 Sep 2016. https://youtu.be/IcnV2wu99mw

[2-8] Williams, Matt. Where is Earth in the Milky Way? Universe Today. 13 Jul 2016. https://www.universetoday.com/65601/where-is-earth-in-the-milky-way/

[2-9] Miller, Heinie. "Battling Frankie Kirk, Lightweight Champ of the Navy, is Slipped a Man Sized Package." *Our Navy, the Standard Publication of the U.S. Navy.* 1914. Volume 8. Page 63.

[2-10] Larson, Cynthia Sue. *Cynthia Sue Larson Interviews Yasunori Nomura.* 3 Dec 2014. https://cynthiasuelarson.wordpress.com/2014/12/03/cynthia-sue-larson-interviews-yasunori-nomura/

[2-11] Parry, Glenn Aparicio. *Original thinking: A radical revisioning of time, humanity, and nature.* North Atlantic Books, 2015.

[2-12] Lachman, Gary. *Lost knowledge of the imagination.* Floris Books, 2017.

[2-13] Gribbin, John. Deep simplicity: Chaos, complexity and the emergence of life. Penguin UK, 2005.

[2-14] Kuhn, Thomas S. The structure of scientific revolutions. University of Chicago press, 2012.

[2-15] Squires, Euan J. *The mystery of the quantum world.* CRC Press, 1994.

CHAPTER THREE
MY MANDELA EFFECT LIFE

[3-1] Brandon, Diane. *Born Aware: Stories & Insights from Those Spiritually Aware Since Birth.* Llewellyn Publications, 2017. Llewellyn Worldwide, 2017.

[3-2] Leiter, L. DAVID. "The vardøgr, perhaps another indicator of the non-locality of consciousness." *Journal of Scientific Exploration* 16, no. 4 (2002): 621-634.

[3-3] Larson, Cynthia. Look Back on Your Past with Love. *RealityShifters.* Issue 129. July 2010.

[3-4] Larson, Cynthia. Seeing Loved Ones Before They Arrive. *RealityShifters.* Issue 133. October 2010. http://realityshifters.com/pages/archives/oct10.html

[3-5] Larson, Cynthia Sue. *Reality Shifts: When Consciousness Changes the Physical World*. RealityShifters, 2012.

[3-6] Harris, Matthew. A Discussion with CEO Chris Anatra: What the Mandela Effect has meant for his business. Geek Insider. 10 Jul 2019. https://www.geekinsider.com/a-discussion-with-ceo-chris-anatra-what-the-mandela-effect-has-meant-for-his-business/

[3-7] Larson, Cynthia Sue. The Top 10 Ways to Shift Reality. *Realityshifters*. 2007. http://realityshifters.com/pages/articles/toptenwaystoshift.html

[3-8] Larson, Cynthia. Healing Things. *Planet Lightworker Magazine*. New Earth Publications. 2002.

[3-9] Proietti, Massimiliano, Alexander Pickston, Francesco Graffitti, Peter Barrow, Dmytro Kundys, Cyril Branciard, Martin Ringbauer, and Alessandro Fedrizzi. "Experimental rejection of observer-independence in the quantum world." *arXiv preprint arXiv:1902.05080* (2019).

[3-10] Larson, Cynthia. Feeling Grateful to be Alive. *RealityShifters*. May 2019. Issue 236. http://www.realityshifters.com/pages/archives/may19.html

[3-11] Brooks, Alexis. "Baffling Mandela Effect: Dead or Alive (or both?)". *Higher Journeys with Alexis Brooks*. May 8 2019. https://youtu.be/dDUUlnfpNDY

[3-12] Yale University Physicists can predict the jumps of Schrodinger's cat (and finally save it). Jun 3, 2019. https://phys.org/news/2019-06-physicists-schrodinger-cat.html

[3-13] Larson, Cynthia Sue. "Be Your Best Self." *RealityShifters*. Feb 2020. Issue 245. http://realityshifters.com/pages/archives/feb20.html

[3-14] Larson, Cynthia Sue. *RealityShifers Guide to High Energy Money*. RealityShifters, 2010.

[3-15] Larson, Cynthia Sue and moneybags73. Art Bell Archives Live—Mandela Monday. Ripon Rabbit Hole YouTube channel. 2 Feb 2020. https://youtu.be/zBhkkW1KTk4

[3-16] Larson, Cynthia Sue. *Aura Advantage: How the Colors in Your Aura Can Help You Attain Your Desires and Attract Success*. Lightworker, 2006.

[3-17] Larson, Cynthia Sue. "Experiencing Alternate Histories." *RealityShifters*. May 2016. Issue 200. http://realityshifters.com/pages/archives/may16.html

CHAPTER FOUR
HISTORY OF THE MANDELA EFFECT

[4-1] Powers, Ron. *Mark Twain: A Life*. Simon and Schuster, 2005.

[4-2] Twain, Mark. "New York Journal." New York, New York, USA. (2 Jun 1897).

[4-3] Jaffe, Aniela. "Memories, dreams, reflections by CG Jung." *Trans. Richard and Clara Winston. New York: Vintage Books* (1989).

[4-4] Wolff, Konrad. "EA Bennett, Meetings with Jung. Conversations recorded during the years 1946-1961. Dairnon Verlag, Zürich 1985. 125 Seiten." *Gesnerus* 43, no. 3-4 (1986): 338-339.

[4-5] Larson, Cynthia Sue. *Reality Shifts: When Consciousness Changes the Physical World*. RealityShifters, 2012.

[4-6] Serling, Rod. "The Parallel." *The Twilight Zone*. Season 4, Episode 11. March 14, 1963.

[4-7] Dick, Philip K. "If you find this world bad, you should see some of the others." *The Shifting Realities of Philip K. Dick: Selected Literary and Philosophical Writings* (1977): 233-258.

[4-8] Peake, Anthony. *A Life of Philip K. Dick: The Man Who Remembered the Future*. Arcturus Publishing, 2013: 135-136.

[4-9] Atwater, P. M. H. *Future memory*. Hampton Roads Publishing, 2013.

[4-10] Tribbe, Frank C. "Research Report: Psychic Leap-Frog by Automobile." Spiritual Frontiers Quarterly Journal, volume 20, number 2. Spring 1988. Philadelphia, Pennsylvania, USA.

[4-11] Art Bell and Cynthia Sue Larson discuss reality shifts and the Mandela Effect. *Midnight in the Desert*. 15 Oct 2015. https://youtu.be/0SMwBx9QrUI

[4-12] Larson, Cynthia Sue. *Quantum Jumps: An Extraordinary Science of Happiness and Prosperity*. RealityShifters, 2013.

[4-13] Larson, Cynthia. Living in Shifting Realities. *RealityShifters*. Issue 70. July 2005. http://realityshifters.com/pages/archives/jul05.html

[4-14] Mandela, Nelson. *In Full: Mandela's Poverty Speech*. BBC. 3 Feb 2005. http://news.bbc.co.uk/2/hi/uk_news/politics/4232603.stm

[4-15] Chadwick, Ed. *New Appeal Over 'Dinosaur': A dinosaur which used to inhabit Bolton Museum and Art Gallery has become 'extinct' for a second time.* "The Bolton News." 23 Jan 2006. https://www.theboltonnews.co.uk/news/676673.new-appeal-over-dinosaur/

[4-16] Starfire Tor website. Http://www.starfiretor.com, begun in 2006.

[4-17] Jones, Marie D., and Larry Flaxman. *This Book is from the Future: A Journey Through Portals, Relativity, Worm Holes, and Other Adventures in Time Travel*. Red Wheel/Weiser, 2012.

[4-18] LeSage, Trish. *Traveling to Parallel Universes*. 2012.

[4-19] Jinks, Tony. *Disappearing object phenomenon: An investigation*. McFarland, 30 Sep 2016.

[4-20] Eriksen, Stasha. The Mandela Effect: Everything is Changing. 11 Jul 2017.

[4-21] Santosusso, Anthony. *Mind Beyond Matter: The Mandela Effect*. 20 May 2018.

[4-22] Shelsky, Rob. *Shattered Reality: The Mandela Effect*. 18 Aug. 2018.

[4-23] Barrington, Mary Rose. *JOTT: When Things Disappear... and Come Back Or Relocate—and why it Really Happens.* Anomalist Books, 21 Sep 2018.

[4-24] Bean, Bill. *"Stranger than Fiction: True Supernatural Encounters of a Spiritual Warrior."* Stellium Books. 2 Dec 2018.

[4-25] DuFay, Dale. *"Terra's of the Milky Way."* 24 Dec 2018.

[4-26] Caladan, Tray S. *Mandela Effect: Analysis of a Worldwide Phenomenon.* 11 Apr 2019.

[4-27] Caladan, Tray S. *Mandela Effect II: More Analysis.* 1 May 2021.

[4-28] Colts, Eileen; Pinto, PauloM.; Robinson, Shane C.; and Vannessa VA. *Mandela Effect: Friend or Foe?* 11:11 Publishing House. Estero, Florida, USA. 2019.

[4-29] Virk, Rizwan. *The Simulated Multiverse: An MIT Computer Scientist Explores Parallel Universes, Quantum Computing, the Simulation Hypothesis and the Mandela Effect.* Bayview Books. 13 Oct 2021.

[4-30] Duran, Anthony J. *Breaking Reality: Inside the Mandela Effect.* 29 Apr 2023.

[4-31] Harmony Mandela Effect. We met in person – DFW Mandela Effect meetup summary. 11 Jun 2017. https://youtu.be/-- hKnPMzSM

[4-32] Vannessa VA. Mandela Effect Conference 2018 Part 1: with Meegs B, SMQ A.I., and John Boyle. Manassas, Virginia, USA. Aug 2018. https://youtu.be/_PSUOvFB62o

[4-33] Vannessa VA. Mandela Effect Conference 2018 Part 2: with Meegs B, SMQ A.I., and John Boyle. Manassas, Virginia, USA. Aug 2018. https://youtu.be/HKUqmWUQJuw

[4-34] Robinson, Shane C. "The Mandela Effect: How messages embedded in changes led me to my life's purpose." International Mandela Effect Conference. Nov 9, 2019. Ketchum, Idaho. https://youtu.be/WhgTfoCdupw

[4-35] Anatra, Chris. "Awakening in the Dream." International Mandela Effect Conference. Nov 8 2019. Ketchum, Idaho. https://youtu.be/GeJtVTduFHY

[4-36] Larson, Cynthia Sue. "The Science and History of Reality Shifts and Mandela Effects." International Mandela Effect Conference. Nov 8, 2019. Ketchum, Idaho. https://youtu.be/3VgXV-pcY0w

[4-37] Squires, Sharon. "A Sociological Examination of the Mandela Effect." Midwestern Mandela Effect Conference. St. Louis, Missouri, USA. 15 Nov 2019. https://youtu.be/ToBfCFwBqF8

[4-38] Bean, Bill. "Significance of the Mandela Effect Bible Changes" Midwestern Mandela Effect Conference. St. Louis, Missouri, USA. 15 Nov 2019. https://youtu.be/qbwsdJoKF-Y

[4-39] "The Mandela Effect" category, *Jeopardy Questions.* https://jeopardyquestions.com/category/the-mandela-effect

[4-40] Matraia, Evan and A.J. 'the Ripon Rabbit.' *Mandela Monthly*. Rippon Rabbit Hole channel, YouTube.

[4-41] Watkins, Eli and Todd, Brian. Former Pentagon UFO official: 'We may not be alone.' CNN Politics. 19 Dec 2017. https://www.cnn.com/2017/12/18/politics/luis-elizondo-ufo-pentagon/index.html

[4-42] Yingling, Marissa E., Yingling, Charlton W., and Bell, Bethany A. "Faculty Perceptions of Unidentified Aerial Phenomena." *Humanities and Social Sciences Communications*. 10, Article 246. 2023.

[4-43] Levy, David Guy and Schlachtenhaufen, Steffen. *The Mandela Effect*. Periscope Entertainment. 2019.

[4-44] MacFarlane, Brian. *Live Mandela Effect Hangouts – 12/2/2019 Special Guests David Guy Levy and Cynthia Sue Larson*. 2 Dec 2019. https://youtu.be/NjMRvR_vsLM

[4-45] Wilson, John. "How to Improve Your Memory." *How to with John Wilson*. HBO documentary TV Series (2020). Season 1, Episode 3. https://www.imdb.com/title/tt10801534/

[4-46] "Mandela Effect." *Robot Chicken*. Season 11, Episode 20. 11 Apr 2022.

[4-47] Koebler, Jason. "Is CERN Causing Collective Mass Delusion by Creating Portals to Alternate Dimensions? An Investigation." *Motherboard: Tech by Vice*. July 20, 2022. https://www.vice.com/en/article/88qg5v/is-cern-causing-mandela-effect-by-creating-portals-to-alternate-dimensions-an-investigation

[4-48] Newcomb, Tim. "The Mandela Effect-And Your False Memories-Are Real, Scientists Confirm in a New Study." *Popular Mechanics*. August 18, 2022. https://www.popularmechanics.com/science/a40849222/mandela-effect-research/

[4-49] Prasad, Deepasri, and Wilma A. Bainbridge. "The Visual Mandela Effect as Evidence for Shared and Specific False Memories Across People." *Psychological Science* (2022): 09567976221108944.

[4-50] Steimer, Sarah. "Study finds widespread false memories of logos and characters, including Mr. Monopoly and Pikachu." University of Chicago News. July 15, 2022. https://news.uchicago.edu/story/visual-mandela-effect-false-memories-psychology-neuroscience-pikachu-mr-monopoly-waldo

CHAPTER FIVE
MANDELA EFFECT THEORIES

[5-1] De Vito, Stefania, Roberto Cubelli, and Sergio Della Sala. "Collective representations elicit widespread individual false memories." *cortex* 45, no. 5 (2009): 686-687.

[5-2] False Memory. *Wikipedia*. 16 Dec 2019. https://en.wikipedia.org/wiki/False_memory

[5-3] Loftus, Elizabeth F., David G. Miller, and Helen J. Burns. "Semantic integration of verbal information into a visual memory." *Journal of experimental psychology: Human learning and memory* 4, no. 1 (1978): 19.

[5-4] Ripon Rabbit Hole Live—Mandela Monthly: Dr. Elizabeth Loftus. Ripon Rabbit Hole. Hosted by A.J. (Ripon Rabbit) and Evan Matraia (moneybags73), with special guest Dr. Elizabeth Loftus. YouTube 5 July 2020. https://youtu.be/5W-8coXLTgI

[5-5] Neisser, Ulric, and Nicole Harsch. "Phantom flashbulbs: False recollections of hearing the news about Challenger." (1992).

[5-6] Walton, Douglas. *Argumentation schemes for presumptive reasoning.* Routledge, 2013.

[5-7] Jinks, Tony. "The Disappearing Object Phenomenon (DOP)." UFO Research (NSW) Incorporated. Ryde Eastwood Leagues Club, NSW, Australia. 3 Jun 2017. https://youtu.be/_F5BRllyX1U

[5-8] Geordie Rose - Quantum Computing: Artificial Intelligence is Here. ideacity. YouTube. 25 Aug 2015. https://youtu.be/PqN_2jDVbOU

[5-9] Bostrom, Nick. "Are we living in a computer simulation?." *The Philosophical Quarterly* 53, no. 211 (2003): 243-255.

[5-10] Virk, Rizwan. *The Simulated Multiverse: An MIT Computer Scientist Explores Parallel Universes, The Simulation Hypothesis, Quantum Computing and the Mandela Effect.* Vol. 2. Bayview Books, LLC, 2021.

[5-11] Wigner, Eugene P. "The unreasonable effectiveness of mathematics in the natural sciences." In *Mathematics and Science*, pp. 291-306. 1990.

[5-12] Tegmark, Max. *Our mathematical universe: My quest for the ultimate nature of reality.* Vintage, 2014.

[5-13] Hoffman, Donald. *The case against reality: Why evolution hid the truth from our eyes.* WW Norton & Company, 2019.

[5-14] Koebler, Jason. "Is CERN Causing Collective Mass Delusino by Creating Portals to Alternate Dimensions? An Investigation." *Vice.* 20 Jul 2022. https://www.vice.com/en/article/88qg5v/is-cern-causing-mandela-effect-by-creating-portals-to-alternate-dimensions-an-investigation

[5-15] Proietti, Massimiliano, Alexander Pickston, Francesco Graffitti, Peter Barrow, Dmytro Kundys, Cyril Branciard, Martin Ringbauer, and Alessandro Fedrizzi. "Experimental test of local observer independence." *Science advances* 5, no. 9 (2019): eaaw9832.

[5-16] Gupta, Sayantan. "Quantum Suicide Thesis with Quantum Immortality." (2017).

[5-17] SMQ AI. *Mandela Effect: We Died in 2012 | Cymatics | Echoes.* 1 Mar 2017. https://youtu.be/RBGcbWJ_dDI

[5-18] van Stockum, Willem Jacob. "IX.—The gravitational field of a distribution of particles rotating about an axis of symmetry." *Proceedings of the Royal Society of Edinburgh* 57 (1938): 135-154.

[5-19] Gödel, Kurt. "An example of a new type of cosmological solutions of Einstein's field equations of gravitation." *Reviews of modern physics* 21, no. 3 (1949): 447.

[5-20] Deutsch, David. "Quantum mechanics near closed timelike lines." *Physical Review D* 44, no. 10 (1991): 3197.

[5-21] Wolf, Fred Alan. *The Yoga of Time Travel: how the mind can defeat time.* Quest Books, 2012.

[5-22] Ball, Philip. "Hawking rewrites history... backwards." *Nature.* 21 Jun 2006. https://www.nature.com/news/2006/060619/full/060619-6.html

[5-23] Larson, Cynthia Sue. "When Consciousness Changes the Physical World." *2012 Creating Your Own Shift.* Shift Awareness Books. 2011.

[5-24] Craw-Goldman, Candace. Quantum Healers. What's New with the Mandela Effect? With Shane Robinson. https://youtu.be/wsajoHbk32E

[5-25] Marsh, Roger. *TruthBubble: Navigating Realms of Reality and Our Societal Shift from Fear to Love.* 2021.

[5-26] Parry, Glenn Aparicio. *Original thinking: A radical revisioning of time, humanity, and nature.* North Atlantic Books, 2015.

[5-27] Alford, Dan Moonhawk. *A Report on the Fetzer Institute-sponsored Dialogues between Western and Indigenous Scientists.* Society for the Anthropology of Consciousness. 11 Apr 1993. http://hilgart.org/enformy/dma-b.htm

[5-28] Dick, Philip K. "If you find this world bad, you should see some of the others." Metz, France. 1977. https://youtu.be/RkaQUZFbJjE

[5-29] Lloyd, Seth, Lorenzo Maccone, Raul Garcia-Patron, Vittorio Giovannetti, Yutaka Shikano, Stefano Pirandola, Lee A. Rozema et al. "Closed timelike curves via postselection: theory and experimental test of consistency." *Physical review letters* 106, no. 4 (2011): 040403.

[5-30] Wolf, Fred Alan. "Ontology, Epistemology, Consciousness; And Closed, Timelike Curves." *Cosmos and History: The Journal of Natural and Social Philosophy* 13, no. 2 (2017): 65-94.

[5-31] Dick, Philip K. "If you find this world bad, you should see some of the others." Metz, France. 1977. https://youtu.be/RkaQUZFbJjE

[5-32] Larson, Cynthia Sue. *Reality Shifts: When Consciousness Changes the Physical World.* RealityShifters, 2012.

[5-33] Meyers, Bryant A. *Pemf-the Fifth Element of Health: Learn Why Pulsed Electromagnetic Field (Pemf) Therapy Supercharges Your Health Like Nothing Else!.* BalboaPress, 2013.

[5-34] Hicks, Jerry. "ME Research: Possible Major Breakthrough." DarkWolf's Den. 15 Apr 2020. https://youtu.be/ngQ7zH4_XUI

[5-35] Larson, C. "Comes True, Being Hoped For; Time of acceptance, time of change." *PARABOLA-NEW YORK*-25, no. 1 (2000): 84-89.

[5-36] Calleman, Carl Johan. *The Nine Waves of Creation: Quantum Physics, Holographic Evolution, and the Destiny of Humanity.* Simon and Schuster, 2016.

[5-37] Anderson, Norman B. et al. "Stress in America: Missing the Health Care Connection." American Psychological Association. 7 Feb 2013.

[5-38] Norton, Amy. "Number of Americans practicing yoga, meditation surged in last six years." Health Day News. 8 Nov 2018. https://consumer.healthday.com/fitness-information-14/yoga-health-news-294/yoga-meditation-surging-in-popularity-in-u-s-739498.html

[5-39] Baer, Drake. "Here's What Google Teaches Employees In Its 'Search Inside Yourself' Course." *Business Insider.* 5 Aug 2014. https://www.businessinsider.com/search-inside-yourself-googles-life-changing-mindfulness-course-2014-8?

[5-40] Pickert, Kate. "The Mindful Revolution: The Science of Finding Focus in a Stressed-Out, Multitasking Culture." *Time.* 3 Feb 2014.

[5-41] Roemigk, Alyssa. "Lotus Pose on Two: The Seahawks believe their kinder, gentler philosophy is the future of football. ESPN The Magazine. 21 Aug 2013.

[5-42] Leibniz, Gottfried Wilhelm. "Principles of Nature and Grace (1714)." (1992)

[5-43] Leibniz, Gottfried Wilhelm, and Gottfried Wilhelm Freiherr von Leibniz. *Leibniz: New essays on human understanding.* Cambridge University Press, 1996.

[5-44] "Meaningful Coincidences with Bernard Beitman." *Living the Quantum Dream with Cynthia Sue Larson.* DreamVisions7 Radio Network. November 5, 2022. https://dreamvisions7radio.com/bernard-beitman/

[5-45] Beitman, Bernard. *Meaningful Coincidences: How and Why Synchronicity and Serendipity Happen.* Park Street Press. 2022

[5-46] Ogburn, William F., and Dorothy Thomas. "Are inventions inevitable? A note on social evolution." *Political science quarterly* 37, no. 1 (1922): 83-98.

[5-47] Robson, David. "Why speaking to yourself in the third person makes you wiser." ScienceBeta. 7 Aug 2019. https://sciencebeta.com/third-person-wisdom/

[5-48] Munroe, Randall. "Brussels Sprouts Mandela Effect." Xkcd. 2019.

CHAPTER SIX
SCIENCE OF THE MANDELA EFFECT

[6-1] Busemeyer, Jerome R., and Peter D. Bruza. *Quantum models of cognition and decision.* Cambridge University Press, 2012.

[6-2] Larson, Cynthia Sue. Life on the Edge with Johnjoe McFadden. *Living the Quantum Dream.* DreamVisions7Radio. 22 Dec 2015. https://dreamvisions7radio.com/life-on-the-edge-with-johnjoe-mcfadden-2/

[6-3] Hoffman, Donald and Larson, Cynthia. Perception, Truth, and Reality with Donald Hoffman. *Living the Quantum Dream.* DreamVisions7Radio

Network. 2 Apr 2016. https://dreamvisions7radio.com/perception-truth-and-reality-with-donald-hoffman/

[6-4] Hoffman, Donald. *The case against reality: Why evolution hid the truth from our eyes*. WW Norton & Company, 2019.

[6-5] "A quantum experiment suggests there's no such thing as objective reality." *MIT Technology Review*. 12 Mar 2019.

[6-6] Larson, Cynthia Sue. "Physics Experiment Challenges Objective Reality. 1 Apr 2019. https://cynthiasuelarson.wordpress.com/2019/04/01/physics-experiment-challenges-objective-reality/

[6-7] Jennings, David, and Matthew Leifer. "No return to classical reality." *Contemporary Physics* 57, no. 1 (2016): 60-82.

[6-8] Thorburn, William. The Myth of Ockham's Razor. Mind 27. 1918. pp. 345-353. http://www.logicmuseum.com/authors/other/mythofockham.htm

[6-9] Larson, Cynthia Sue. "Primacy of quantum logic in the natural world." *Cosmos and History: The Journal of Natural and Social Philosophy* 11, no. 2 (2015): 326-340.

[6-10] Cairns, John, Julie Overbaugh, and Stephan Miller. "The origin of mutants." *Nature* 335, no. 6186 (1988): 142-145.

[6-11] McFadden, Johnjoe, and Jim Al-Khalili. *Life on the edge: the coming of age of quantum biology*. Broadway Books, 2016.

[6-12] Scholes, Gregory D. "Quantum-coherent electronic energy transfer: Did nature think of it first?." *The Journal of Physical Chemistry Letters* 1, no. 1 (2010): 2-8.

[6-13] Eugene, P. "Wigner. Remarks on the mind-body question." (1961).

[6-14] Guérin, Philippe Allard, Veronika Baumann, Flavio Del Santo, and Časlav Brukner. "A no-go theorem for the persistent reality of Wigner's friend's perception." *arXiv preprint arXiv:2009.09499* (2020).

[6-15] Linder, Courtney. "It's Impossible to Tell if this Story Exists According to Quantum Physics: Blame it on the Wigner's Friend paradox. *Popular Mechanics*. 23 Aug 2021.

[6-16] Cavalcanti, Eric G. "The view from a Wigner bubble." *Foundations of Physics* 51, no. 2 (2021): 39.

[6-17] Weissmann, George, and Cynthia Sue Larson. "The quantum paradigm and challenging the objectivity assumption." *Cosmos and History: The Journal of Natural and Social Philosophy* 13, no. 2 (2017): 281-297.

[6-18] Zych, Magdalena, Fabio Costa, Igor Pikovski, and Časlav Brukner. "Bell's theorem for temporal order." *Nature communications* 10, no. 1 (2019): 1-10.

[6-19] Fein, Yaakov Y., Philipp Geyer, Patrick Zwick, Filip Kiałka, Sebastian Pedalino, Marcel Mayor, Stefan Gerlich, and Markus Arndt. "Quantum superposition of molecules beyond 25 kDa." *Nature Physics* (2019): 1-4.

[6-20] Gribbin, John. *Deep simplicity: Chaos, complexity and the emergence of life.* Penguin UK, 2005.

[6-21] Reich, Eugenie Samuel, "Quantum Paradox Seen in Diamond," Nature, 20 August 2013

[6-22] Patil, Yogesh Sharad, Srivatsan Chakram, and Mukund Vengalattore. "Quantum Control by Imaging: The Zeno effect in an ultracold lattice gas." *arXiv preprint arXiv:1411.2678* (2014).

[6-23] Larson, Cynthia. *Reality Shifts: When Consciousness Changes the Physical World.* 2011.

[6-24] Sudarshan, E.C.G. (1983). Perception of quantum systems. In Old and New Questions in Physics, Cosmology, Philosophy, and Theoretical Biology, ed. by A. van der Merwe, Plenum, New York, pp. 457–467.

[6-25] Atmanspacher, Harald, and Thomas Filk. "The Necker–Zeno Model for Bistable Perception." *Topics in cognitive science* 5.4 (2013): 800-817.

[6-26] Riddle, Justin and Larson, Cynthia. "How to Break Cell Phone Addiction Using Quantum Zeno Effect." *Foundations of Mind.* 1 Jul 2015 https://youtu.be/siJRYT1rbEU

[6-27] Taylor, John. *Superminds: A Scientist Looks at the Paranormal.* Viking Adult. 1975.

[6-28] Bell, Philip. "Hawking rewrites history... backwards." *Nature.* (2006).

[6-29] Hawking, Stephen W., and Thomas Hertog. "Populating the landscape: A top-down approach." *Physical Review D* 73, no. 12 (2006): 123527.

[6-30] Bohm Dialogue: A radically new vision of dialogue https://www.bohmdialogue.org/

[6-31] Wolf, Fred Alan. "Ontology, Epistemology, Consciousness; And Closed, Timelike Curves." *Cosmos and History: The Journal of Natural and Social Philosophy* 13, no. 2 (2017): 65-94.

[6-32] Larson, Cynthia. "Evidence of shared aspects of complexity science and quantum phenomena." *Cosmos and History: The Journal of Natural and Social Philosophy* 12, no. 2 (2016): 160-171.

[6-33] Dong, Ming-Xin, Dong-Sheng Ding, Yi-Chen Yu, Ying-Hao Ye, Wei-Hang Zhang, En-Ze Li, Lei Zeng et al. "Temporal Wheeler's delayed-choice experiment based on cold atomic quantum memory." *npj Quantum Information* 6, no. 1 (2020): 1-7.

[6-34] Backster, Cleve. *Primary perception: Biocommunication with plants, living foods, and human cells.* White Rose Millennium Press, 2003.

[6-35] Sheldrake, Rupert. *Dogs that know when their owners are coming home: And other unexplained powers of animals.* Broadway Books, 2011.

[6-36] Freeman, Walter J. *Societies of brains: A study in the neuroscience of love and hate.* Psychology Press, 2014.

[6-37] Frauchiger, Daniela, and Renato Renner. "Quantum theory cannot consistently describe the use of itself." *Nature communications* 9, no. 1 (2018): 1-10.

[6-38] Evans, Peter W. "Perspectival objectivity." *European Journal for Philosophy of Science* 10, no. 2 (2020): 1-21.

[6-39] Fuchs, Christopher A., and Blake C. Stacey. "QBism: Quantum theory as a hero's handbook." In *Proceedings of the International School of Physics "Enrico Fermi*, vol. 197, pp. 133-202. 2019.

[6-40] McFadden, Johnjoe, and Jim Al-Khalili. *Life on the edge: the coming of age of quantum biology.* Broadway Books, 2016.

[6-41] Wang, Zheng, et al. "The potential of using quantum theory to build models of cognition." *Topics in Cognitive Science* 5.4 (2013): 672-688.

[6-42] Bowles, Joseph, Tamás Vértesi, Marco Túlio Quintino, and Nicolas Brunner. "One-way einstein-podolsky-rosen steering." *Physical Review Letters* 112, no. 20 (2014): 200402.

[6-43] Sainz, Ana Belén, Leandro Aolita, Nicolas Brunner, Rodrigo Gallego, and Paul Skrzypczyk. "Classical communication cost of quantum steering." *Physical Review A* 94, no. 1 (2016): 012308.

[6-44] Turin, Luca. "A spectroscopic mechanism for primary olfactory reception." *Chemical senses* 21, no. 6 (1996): 773-791.

[6-45] Proietti, Massimiliano, Alexander Pickston, Francesco Graffitti, Peter Barrow, Dmytro Kundys, Cyril Branciard, Martin Ringbauer, and Alessandro Fedrizzi. "Experimental rejection of observer-independence in the quantum world." *arXiv preprint arXiv:1902.05080* (2019).

[6-46] Cramer, John G. "The quantum handshake." *NY: Springer Publishing Co* (2016).

CHAPTER SEVEN
MANDELA EFFECT EXPERIENCERS

[7-1] Larson, Cynthia. "How Do You Shift Reality? The First Ever Reality Shifting Survey Results Are In! *RealityShifters.* April 2000. http://realityshifters.com/pages/research/apr2000.html

[7-2] Lupo, Tarrin P. Mandela Effect – Poll #1-Why Us? What Do We All Have In Common? 2017. *Dr. Tarrin P. Lupo.* 10 Jan 2017. https://youtu.be/wUA26nN7DHs

[7-3] Briggs, Katharine Cook, and Isabel Briggs Myers. *The Myers-Briggs Type Indicator: Form G.* Consulting Psychologists Press, 1977.

[7-4] Keirsey, David, and Marilyn M. Bates. *Please understand me.* Prometheas Nemesis, 1984.

[7-5] How Frequent is my type. Myers & Briggs Foundation. https://www.myersbriggs.org/my-mbti-personality-type/my-mbti-results/how-frequent-is-my-type.htm?bhcp=1

[7-6] Larson, Cynthia. *Quantum Jumps.* 2013

[7-7] Beitman, Bernard. *Meaningful Coincidences: How and Why Synchronicity and Serendipity Happen.* Park Street Press. 2022

[7-8] Sovereign Sage, The Mr. Smith Effect (Escape the Matrix). 7 Dec 2017. https://youtu.be/2myLcH9oRz8

[7-9] Robertson, Morgan. *Futility.* Vol. 3, no. 4577. MF Mansfield, 1898.

[7-10] Clark, Nick. "How Homer Simpson discovered the Higgs boson over a decade before scientists." *Independent.* March 1, 2015. https://www.independent.co.uk/news/science/how-homer-simpson-discovered-the-higgs-boson-over-a-decade-before-scientists-10079006.html

[7-11] French, Aaron. "The Mandela Effect and New Memory." *Correspondences* 6, no. 2 (2019).

[7-12] Larson, Cynthia. "If Artificial Intelligence Asks Questions, Will Nature Answer? Preserving Free Will in a Recursive Self-Improving Cyber-Secure Quantum Computing World." *Cosmos and History: The Journal of Natural and Social Philosophy* 14, no. 1 (2018): 71-82.

[7-13] Lachman, Gary. *Lost knowledge of the imagination.* Floris Books, 2017.

[7-14] Calleman, Carl Johan. *The Nine Waves of Creation: Quantum Physics, Holographic Evolution, and the Destiny of Humanity.* Simon and Schuster, 2016.

[7-15] Badzey, Robert L., and Pritiraj Mohanty. "Coherent signal amplification in bistable nanomechanical oscillators by stochastic resonance." *Nature* 437, no. 7061 (2005): 995-998.

[7-16] "Taming the Multiverse—Stephen Hawking's Final Theory About the Big Bang." Phys.org News. 2 May 2018. University of Cambridge. https://phys.org/news/2018-05-multiversestephen-hawking-theory-big.html

[7-17 Sheldrake, Rupert. *The presence of the past: Morphic resonance and the habits of nature.* Icon Books Ltd, 2011.

[7-18] Freeman, Walter J. *Societies of brains: A study in the neuroscience of love and hate.* Psychology Press, 2014.

[7-19] Lupo, Tarrin P. Positive & Negative Side Effects of the Mandela Effect – New 2017. *Dr. Tarrin P. Lupo.* 9 Jan 2017. https://youtu.be/-5l3Ct_xCK4

[7-20] Larson, Cynthia Sue. *Reality Shifts: When Consciousness Changes the Physical World.* RealityShifters, 2012.

[7-21] HumanMetrics, "Jung Typology Test" personality test. http://www.humanmetrics.com/cgi-win/jtypes2.asp

[7-22] Official Myers-Briggs Instrument. The Myers and Briggs Foundation. https://www.myersbriggs.org/my-mbti-personality-type/take-the-mbti-instrument/

[7-23] Empathy Quiz. UC Berkeley. Greater Good Science Center. https://greatergood.berkeley.edu/quizzes/take_quiz/empathy

[8-1] Larson, Cynthia. *Reality Shifts: When Consciousness Changes the Physical World*

[8-2] White, Rhea. "Exceptional human experiences: A brief overview." *EHE Network, Inc* (1999).

[8-3] Singer, Michael. *The untethered soul: The journey beyond yourself.* New Harbinger Publications, 2007.

[8-4] Moorjani, Anita. *Dying to be me: My journey from cancer, to near death, to true healing.* Hay House, Inc, 2022.

[8-5] Radin, Dean I. *Supernormal: Science, yoga, and the evidence for extraordinary psychic abilities.* Deepak Chopra, 2013.

[8-6] Schmidt, Helmut. "PK effect on pre-recorded targets." *Journal of the American Society for Psychical Research* 70, no. 3 (1976): 267-291.

[8-7] Schmidt, Helmut, and Henry Stapp. "PK with prerecorded random events and the effects of preobservation." *The Journal of Parapsychology* 57, no. 4 (1993): 331-350.

[8-8] Documentary "Ring of Fire" and also by Catherine Cooke from Mind Science Foundation, Dr. Roger Nilson, and Dr. Gregory Simpson from Albert Einstein College of Medicine

[8-9] Gough, William C., Larson, Cynthia, Hiersch, Richard, Martinez, Bett. Notes on Conversation with a Shaman from Mongolia. Foundation for Mind-Being Research. 6 Sep 2004.

[8-10] Cameron, Grant and Barnabe, Desta. *Contact Modalities: The Keys to the Universe.* 2020.

[8-11] Barrington, Mary Rose. *JOTT: When Things Disappear... and Come Back Or Relocate—and why it Really Happens.* Anomalist Books, 2018.

[8-12] Larson, Cynthia Sue. "Felt with the Heart." *RealityShifters.* Aug 2008. Issue 107. http://realityshifters.com/pages/archives/aug08.html

[8-13] O Nuallain, Sean. "The Practice of Presence; Consciousness, Meditation, Health and Spirituality." *Cosmos and History: The Journal of Natural and Social Philosophy* 14, no. 2 (2018): 178-206.

[8-14] Larson, C. "Comes True, Being Hoped For; Time of acceptance, time of change." *PARABOLA-NEW YORK*-25, no. 1 (2000): 84-89.

[8-15] Martineau, LaVan. *The rocks begin to speak.* Las Vegas, Nev.: KC Publications, 1973.

[8-16] Deshazer, William. Back from the dead. Nature Conservancy. Winter 2019. pp. 9-10.

[8-17] Nelson, Bryan. Lazarus species: 13 'extinct' animals found alive. MNN. https://www.mnn.com/earth-matters/animals/photos/lazarus-species-13-extinct-animals-found-alive/rediscovered

[8-18] Holland, Jennifer S. Once thought extinct, bizarre horned frog reappears in Ecuador. National Geographic. 12 Dec 2018. https://www.nationalgeographic.com/animals/2018/12/lost-marsupial-frog-rediscovered-ecuador-choco-forest/

[8-19] Langlois, Jill. How an 'extinct' tortoise was rediscovered after a century. National Geographic. 22 Feb 2019. https://www.nationalgeographic.com/animals/2019/02/extinct-fernandina-giant-tortoise-found/

[8-20] Jensen, Evelyn L., Stephen J. Gaughran, Nicole A. Fusco, Nikos Poulakakis, Washington Tapia, Christian Sevilla, Jeffreys Málaga, Carol Mariani, James P. Gibbs, and Adalgisa Caccone. "The Galapagos giant tortoise Chelonoidis phantasticus is not extinct." *Communications Biology* 5, no. 1 (2022): 546.

[8-21] Coelacanths. National Geographic. https://www.nationalgeographic.com/animals/fish/group/coelacanths/

[8-22] Nelson, Bryan. Lazarus species: 13 'extinct' animals found alive. MNN. https://www.mnn.com/earth-matters/animals/photos/lazarus-species-13-extinct-animals-found-alive/rediscovered

[8-23] "Rare Clam Thought to be extinct—found alive." *SciTechDaily*. U.C. Santa Barbara. 29 Dec 2022. https://scitechdaily.com/rare-clam-thought-to-be-extinct-found-alive/

[8-24] Laidlaw, Clint. "I didn't know these whales existed, and I'm a Zoologist." *Clint's Reptiles*. 13 Aug 2022. https://youtu.be/vefi6hciYME

[8-25] Parry, Glenn Aparicio. *Original thinking: A radical revisioning of time, humanity, and nature*. North Atlantic Books, 2015.

[8-26] Pirandello, Luigi. *One, No one and one hundred thousand*. Vol. 18. Ravenio Books, 1990.

[8-27] Fadiman, James, and Gruber, Jordan. *Your symphony of selves*. New York: Simon & Schuster, 2020.

[8-28] Dovey, Dana. "Blind woman with dissociative identity disorder spontaneously regains vision in a few of her 10 personalities." *Medical Daily*. 30 Nov 2015.

[8-29] Goleman, Daniel. "Probing the Enigma of Multiple Personality." *The New York Times*. 28 Jun 1988. Section C, Page 1.

[8-30] Wilkinson, William Cleaver. *Classic German Course in English*. Vol. 6. Chautauqua Press, 1887.

[8-31] Lupoff, Richard A. "A Conversation with Philip K. Dick." *Science Fiction Eye* 1, no. 2 (1984).

CHAPTER NINE
ACCIDENTAL HERO'S JOURNEY & GIFTS

[9-1] Campbell, Joseph, and Bill Moyers. *The power of myth*. Anchor, 2011.

[9-2] Moyers, Bill. Mythology of Star Wars. 18 Jun 1999. https://billmoyers.com/content/mythology-of-star-wars-george-lucas/

[9-3] Singer, Michael. *The untethered soul: The journey beyond yourself*. New Harbinger Publications, 2007.

[9-4] Rescher, Nicholas. *Axiogenesis: An essay in metaphysical optimalism*. Lexington Books, 2010.

[9-5] Schrödinger, Erwin. *What is life? The physical aspect of the living cell*. At the University Press, 1951.

CHAPTER TEN
HUMANITY'S GREAT AWAKENING

[10-1] Fitzgerald, Madeline. In a Historic Finale, 8 Spellers Just Won the 2019 Scripps National Spelling Bee. Time.

[10-2] Schwartz, Stephan A. *The 8 laws of change: How to be an agent of personal and social transformation*. Simon and Schuster, 2015.

[10-3] Sarris, Greg. *Mabel McKay: Weaving the Dream*. Vol. 1. Univ of California Press, 1994.

[10-4] Wolf, Jessica. "Alumnus Returns to UCLA to deliver Regents Lecture on the future of Indian Country." *UCLA Newsroom*. 10 Oct 2022. https://newsroom.ucla.edu/stories/greg-sarris-regents-lecture-indian-country

[10-5] Martineau, LaVan. *The rocks begin to speak*. Las Vegas, Nev.: KC Publications, 1973.

[10-6] Larson, C. S. "Comes true, being hoped for-The Hopi understanding of how things change." *PARABOLA-MYTH TRADITION AND THE SEARCH FOR MEANING* 25, no. 1 (2000): 84-87.

[10-7] Miqlos. Project Camelot, Bill Wood, Above & Beyond Project Looking Glass. Jan 2012. https://www.bitchute.com/video/x89kaiHX36vR

[10-8] Sychev, Demid V., Alexander E. Ulanov, Anastasia A. Pushkina, Matthew W. Richards, Ilya A. Fedorov, and Alexander I. Lvovsky. "Enlargement of optical Schrödinger's cat states." *Nature Photonics* 11, no. 6 (2017): 379.

[10-9] Cavalcanti, Eric G. "The view from a Wigner bubble." *arXiv preprint arXiv:2008.05100* (2020).

[10-10] Dummett, A. E., and Antony Flew. "Symposium: Can an effect precede its cause?." *Proceedings of the Aristotelian Society, Supplementary Volumes* (1954): 27-62.

[10-11] Merali, Zeeya. "Back from the future." *Discover Magazine August* 26 (2010).

[10-12] Schmidt, Helmut, and Henry Stapp. "PK with prerecorded random events and the effects of preobservation." *The Journal of Parapsychology* 57, no. 4 (1993): 331-350.

[10-13] Dunne, Brenda J., and Robert G. Jahn. *Consciousness and Anomalous Phisycal Phenomena*. Princeton University, 1995.

[10-14] Leibovici, Leonard. "Effects of remote, retroactive intercessory prayer on outcomes in patients with bloodstream infection: randomised controlled trial." *Bmj* 323, no. 7327 (2001): 1450-1451.

[10-15] Ende, Michael. "Momo, or The Strange Story of the Time Thieves." (1984).

[10-16] Levine, Robert N. *A geography of time: On tempo, culture, and the pace of life*. Basic Books, 2008.

[10-17] Rescher, Nicholas. *Axiogenesis: An essay in metaphysical optimalism*. Lexington Books, 2010.

[10-18] Abbot, Edwin. "Flatland." (1884).

[10-19] Leibniz, G.W. "Principles of Nature and Grace." *The Philosophical Works of Leibniz*. New Haven: Tuttle, Morehouse & Taylor. 1890.

[10-20] Larson, Cynthia. Experience Higher Consciousness with Sacred Humility. RealityShifters. Issue 174. March 2014.
http://realityshifters.com/pages/archives/mar14.html

[10-21] Carroll, Robert, and Stephen Prickett, eds. *The Bible: Authorized King James Version*. Oxford Paperbacks, 2008.

[10-22] Forrest, Lynne. *Guiding Principles for Life Beyond Victim Consciousness*. Conscious Living Media. 2011.

[10-23] Larson, Cynthia Sue. Life on the Edge with Lynne Forrest. *Living the Quantum Dream*. DreamVisions7Radio. 27 Dec 2019.
https://dreamvisions7radio.com/life-beyond-victim-consciousness-with-lynne-forrest/

[10-24] Unbiased & On the Fence. Crazy ME's and synchronicities on ME Road Trip. 18 Nov 2019. https://youtu.be/BrgfkvttgSA

[10-25] Alford, Dan Moonhawk. "A report on the Fetzer Institute-sponsored dialogues between Western and indigenous scientists." In *A Presentation for the Annual Spring Meeting of the Society for the Anthropology of Consciousness*. 1993.

[10-26] Bear, Leroy Little. "Jagged worldviews colliding." *Reclaiming Indigenous voice and vision* 77 (2000).

[10-27] Freeman, Walter J. *Societies of brains: A study in the neuroscience of love and hate*. Psychology Press, 2014.

[10-28] Dossey, Larry. *One mind*. Hay House, Inc, 2013.

[10-29] Denworth, Lydia. "Brain Waves Synchronize when People Interact." *Scientific American*. Jul 1, 2023.
https://www.scientificamerican.com/article/brain-waves-synchronize-when-people-interact/

[11-1] Hicks, Jerry. West Coast Mandela Effect Conference Community Panel. 24 Nov 2019. https://youtu.be/qIzUr66gr7U

[11-2] The Mandela Effect in Brazil. International Mandela Effect Conference (IMEC) Open Tables. Season 1, Episode 4. 26 May 2021. https://www.youtube.com/live/lCzm5psm3Yo

[11-3] Parry, Glenn Aparicio. *Original thinking: A radical revisioning of time, humanity, and nature.* North Atlantic Books, 2015.

MANDELA EFFECT RESOURCES

Mandela Effect Books

Barrington, Mary Rose. *JOTT: When Things Disappear... and come back or relocate—and why it really happens.* Anomalist Books. 21 Sep 2018.

Bean, Bill. *"Stranger than Fiction: True Supernatural Encounters of a Spiritual Warrior."* Stellium Books. Dec 2018.

Broome, Fiona. *The Mandela Effect: Theories and Explanations.* Jan 2020.

Caladan, Tray S. *Mandela Effect: Analysis of a Worldwide Phenomenon.* 11 Apr 2019.

Colts, Eileen; Pinto, PauloM.; Robinson, Shane C.; and Vannessa VA. *Mandela Effect: Friend or Foe?* 11:11 Publishing House. Estero, Florida, USA. 2019.

DuFay, Dale. *Terra's of the Milky Way.* 2018

Duran, Anthony J. *Breaking Reality: Inside the Mandela Effect.* 29 Apr 2023.

Eriksen, Stasha. *The Mandela Effect: Everything is Changing.* 11 Jul 2017.

Jinks, Tony. *Disappearing Object Phenomenon.* McFarland & Company. 14 Oct 2016.

Larson, Cynthia Sue. *Reality Shifts: When Consciousness Changes the Physical World.* RealityShifters, 2012.

Larson, Cynthia Sue. *Quantum Jumps: An Extraordinary Science of Happiness and Prosperity.* RealityShifters, 2013.

LeSage, Trish. *Traveling to Parallel Universes.* 2012.

Santosusso, Anthony. *Mind Beyond Matter: The Mandela Effect.* 2018.

Shelsky, Rob. *Shattered Reality: The Mandela Effect.* 18 Aug. 2018.

Virk, Rizwan. *The Simulated Multiverse: An MIT Computer Scientist Explores Parallel Universes, The Simulation Hypothesis, Quantum Computing and the Mandela Effect.* Vol. 2. Bayview Books, LLC, 2021.

Mandela Effect Articles

French, Aaron. "The Mandela effect and new memory." *Correspondences* 6, no. 2 (2019).

Prasad, Deepasri, and Wilma A. Bainbridge. "The Visual Mandela Effect as evidence for shared and specific false memories across people." *Psychological Science* 33, no. 12 (2022): 1971-1988.

First-Hand Mandela Effects Accounts

Fiona Broome's Mandela Effect Site
https://mandelaeffect.com/major-memories

Cynthia Sue Larson's RealityShifters Site
http://realityshifers.com/pages/yourstories.html

Mandela Effect Blogs and Websites

Alternate Memories
https://www.alternatememories.com/

Mandela Effect Database
https://www.flickr.com/photos/154930084@N08/albums

Mandela Effect
https://mandelaeffect.com/

Mandela Effect Bible Changes
http://mandelabiblechanges.com/

Mandela Effect Proof
http://mandelaeffectproof.blogspot.com/

RealityShifters
http://realityshifters.com/

Timeline Shift
https://www.timelineshift.com/

Mandela Effect Conferences

International Mandela Effect Conference (IMEC)
www.imec.world

Mandela Effect Ezines

RealityShifters
www.RealityShifters.com/pages/news.html

FaceBook

Mandela Effect
Mandela Effect Investigators
Mandela Effect Society
Mandela Reality Residue Hunters
One Mind, Many Worlds

Carter Tweed's Mandela Effect Test

https://www.alternatememories.com/

Reddit

/r/MandelaEffect
/r/Retconned
/r/Mandela_Effect

/r/Glitch_In_The_Matrix
/r/GlitchInTheMatrix

YouTube Channels

AffectedCollective
Aggroed Lighthacker
Armor Up!
Basketofcups
Brian MacFarlane
Brian Stavely
Carla RedPill
Changing Matrix
Crazies Lab
Cynthia Sue Larson
DarkWolf's Den
DigitalScribe
Dr. Tarrin P. Lupo
EMAN31
Evin Powers
Gemini Vision
God's Armor
Guy Fauqes
Harmony Mandela Effect
Harold Beebe
Hidden Knowledge
I Am Lazlow
International Mandela Effect Conference (IMEC)
J W
Jacob Israel
Jynx Cat
Kiera
The Kev Baker Show
LIFE MATRIX

Lone Eagle
Mandela Affected
Marlena Effect
Meegs B
Michelle Platti
Moneybags73
Nathiel Winter-Hebert
NoblenessDee
Once Upon a Timeline
OpenYourReality
Photohelix
Psionic League
Quantum Businessman
Rain on the Plains
RemovingtheShackles
Rena Jacile
Ripon Rabbit Hole
Scarabperformance
Scott Harrison
Skook's Strange World
SMQ AI
SoulSpaUniversity
Susso
System Failure
Unbiased & On the Fence
Vannessa VA
Wakeuporelse PMA
William Thorg

INDEX

Curious George, 97

D

da Vinci, Leonardo, 55, 238
Darth Vader, 45
Dazed and Confused movie, 100
de Chardin, Teilhard, 131
déjà vu, 48, 55, 61, 123, 155, 184, 198, 235, 259-260
DeMario, Nicole, 13, 107
Denmark, 27-28, 79, 105, 255
Dick, Philip K. (PKD), 92-93, 103, 117, 125-128, 206
dinosaur, 47, 96-97, 102, 255
deep listening, 243-245, 249, 252
delayed choice, 123, 148, 155-156, 159, 199, 224
dimensional shift, 48, 55
disappearances, 98, 101
Disney, Walt, 15, 189
distorted memory, 114
do-overs, 48, 58
double memories, 48, 49
double slit experiment, 147-148, 159, 239
double-stuffed Oreos, 31, 120
download, 48-50, 57-59, 154, 261
drama triangle, 241
dreams, 43-44, 48, 52, 93, 105, 128, 153, 186, 195, 203-204, 207, 247-248
DuFay, Dale, 13, 102, 284
Dunne, Brenda, 236
Dunne, John William, 55
Duran, Anthony, 13, 103, 284
D-Wave, 53, 117

E

empaths, 172, 174, 176-177, 179, 258
Ende, Michael, 236
energy states, 220, 223
Engels-Smith, Jan, 13, 107
England, 17, 68, 96-97, 255
entanglement, 120, 133, 139, 153-155, 160, 163, 200, 220, 223, 234, 256
EPR steering, 153, 156, 160, 164, 220, 225
Eriksen, Stasha, 13, 99, 284
Etiäinen, 67

Everything, Everywhere, All at Once movie, 222
evolution of consciousness, 113, 124, 129, 133, 172, 232
Exceptional Human Experience (EHE), 29, 49, 63, 190, 194, 196, 204, 216
explicate order, 160, 200
extinct, 45, 52-53, 117, 193, 201-203, 219, 260

F

Fadiman, James, 205
false memories, 15, 17, 30, 99, 107, 110, 113-114, 116, 129, 189-190
Fauqes, Guy, 36, 103-104, 286
Fifth World, 49-51, 201
Finland, 67
flashbulb memories,50, 115
flat tire fixed by God, 75-77
Flatland, 54, 239
flip-flop, 22, 124, 212, 256
"Fly, my pretties", 115
flyaways, 51, 100-101, 111
Forrest, Lynne, 241
France, 27, 92-93, 125
free will, 50, 55, 58, 225, 232, 248, 263
Freeman, Walter J., 13, 160, 185, 246
French, Aaron, 33, 180
Friends TV show, 100
Fringe TV show, 124
future memory, 18, 50, 56, 93, 123, 155, 184, 198, 235
future self, 66, 156-157, 206-207, 209, 235, 245, 247, 252

G

Gaelic, 181, 199, 200, 224
Galápagos, 202
Geller, Uri, 74, 94, 152
Germany, 27, 125, 205
Gibraltar, 37
Gilligan's Island TV show, 105, 256
glitch in the matrix, 16, 36, 50, 52, 57, 60
God, 54, 75, 77, 104, 120, 126-127, 131, 169, 179, 204, 250, 253
Goddard, Neville, 41, 58
Goethe, Johann Wolfgang von, 206
Golden Age, 49-50
"Good night, Gracie", 115

ABOUT THE AUTHOR

Cynthia Sue Larson is a best-selling author of several books including *Quantum Jumps*, *Reality Shifts*, and *High Energy Money*. Cynthia has a degree in physics from UC Berkeley, an MBA degree, a Doctor of Divinity, and a second degree black belt in Kuk Sool Won. Cynthia is the founder of RealityShifters, first president of the International Mandela Effect Conference, managing director of Foundations of Mind, and creator and host of *Living the Quantum Dream* podcast. She has been featured in numerous shows including Gaia, the History Channel, Coast to Coast AM, One World with Deepak Chopra, and BBC. Cynthia reminds us to ask in every situation, "How good can it get?" Subscribe to her free monthly ezine at: www.realityshifters.com

The Mandela Effect and its Society:
Awakening from Me to We

Could immediate evolutionary improvements, instantaneous geographic and anatomical changes, and miraculous spontaneous healing—including recovery from death—be providing us with clues to essential technology, just when we need it most? At this time when a phenomenon of collective alternate memories known as the *Mandela Effect* is gaining popular interest, people are noticing changes to: movies, books, music, food, clothing, animals, anatomy, geography, astronomy, history, and every aspect of life. With people sharing specifically different memories than official history describes, many questions now arise. Why is the Mandela Effect happening? What do experiencers of the Mandela Effect have in common? Do versions of us exist in parallel universes? How can we know if we are experiencing different timelines? Where is the Mandela Effect taking us?

Cynthia Sue Larson describes how the Mandela Effect is much more than "false collective memory," providing a rich tapestry of historical and scientific background as to how this phenomenon is observed. Answers to questions about the Mandela Effect are provided via insights from indigenous wisdom, quantum science, and careful review of first-hand case studies of Mandela Effects. This examination of what may be the most important topic of our times culminates in exploration of practical techniques for harnessing our natural quantum superpowers at this pivotal time of expanding human consciousness.